Acclaim for

THE GREAT QUAKE

"In his entertaining and enlightening book *The Great Quake*, Henry Fountain tells the story of what is sometimes called 'Alaska's Good Friday Earthquake.' . . . Fountain sets the scene for an abrupt wake-up call, and his description of how it unfolds is gripping."

—*San Francisco Chronicle*

"*The Great Quake* explains how one of North America's worst recent natural disasters led to a fascinating insight. Henry Fountain offers a gripping tale of loss, heroism and, ultimately, discovery."

—Elizabeth Kolbert, Pulitzer Prize–winning author of *The Sixth Extinction*

"As elegant as the score of a Beethoven symphony . . . journalist Henry Fountain provides us with a forthright and timely reminder of the startling historical consequences of North America's largest known earthquake."

—*Nature*

"Henry Fountain knows earthquakes, and he knows how to spin a yarn. *The Great Quake* is the fascinating result. It takes meticulous research and real narrative skill to tell a story that moves this fast yet still feels so complete. The book shines on two levels: as a portrait of two quirky frontier communities before, during and after a stunning disaster, and as the story of an unpretentious geologist whose brilliant analysis of the great quake's causes provided crucial backing for one of the biggest ideas in all of science."

—Dan Fagin, Pulitzer Prize–winning author of *Toms River*

"*The Great Quake* is an elegant showcase of how the progressive work of numerous scientists over time . . . can be pieced together into an idea that reshapes how we see and understand the planet."

—*Science News*

"For five terrifying minutes in 1964, the earth shook beneath Anchorage, Alaska. It devastated the city and towns and villages throughout the state. In this fast-paced, engaging account of that disaster, Henry Fountain tells us what it was like to be there. His interviews with fortunate survivors bear witness to the pluck and determination of the human spirit—and reveal the better side of our nature in times of crisis. Read this book to better understand nature's power—and our human resilience. Fountain's riveting, 'you were there' account pulls you in and keeps you turning the pages to find out who survived—and how."

—Virginia Morell, author
of the *New York Times* bestseller *Animal Wise*

"The detective work involved in reconstructing land movements produced by an earthquake is itself a compelling tale. . . . The book engagingly recounts life in the immediate aftermath of the earthquake."

—*Science*

"Riveting. Science journalism at its best—lucid, clear, engaging and authoritative. My hands were shaking after reading his description of the havoc and raw fury unleashed by Mother Nature in a 9.0 earthquake."

—Michio Kaku, professor of theoretical physics and
New York Times bestselling author of *The Future of the Mind*

"Fountain's new book is a powerful lesson that the term 'solid ground' is one of humanity's greatest illusions. . . . The strength of [*The Great Quake*] is that the veteran science reporter balances anecdotes with a clear explanation of the technical details and what has been learned during the past half-century."

—*Anchorage Daily News*

"Fountain atmospherically depicts life in the frontier communities . . . that were razed when 'the earth (rang) like a bell' for five minutes. . . . The narrative is haunted by images that live long in the mind, not least a crimson tide of dead red snapper flushed from the roiling depths."

—*The Oregonian*

THE GREAT QUAKE

HOW THE BIGGEST EARTHQUAKE IN NORTH AMERICA CHANGED OUR UNDERSTANDING OF THE PLANET

HENRY FOUNTAIN

B\D\W\Y
BROADWAY BOOKS
NEW YORK

Published in the United States by Broadway Books, an imprint of the Crown Publishing Group, a division of Penguin Random House LLC, New York.
crownpublishing.com

Broadway Books and its logo, B \ D \ W \ Y, are trademarks of Penguin Random House LLC.

Originally published in hardcover in the United States by Crown, an imprint of the Crown Publishing Group, a division of Penguin Random House LLC, New York, in 2017.

Library of Congress Cataloging-in-Publication Data is available upon request.

ISBN 978-1-101-90408-4
Ebook ISBN 978-1-101-90407-7

Printed in the United States of America

Maps by Mapping Specialists
Cover design by Michael Morris
Cover photography: Associated Press/AP Photo

10 9 8 7 6 5 4 3 2 1

First Paperback Edition

To George and Doris

CONTENTS

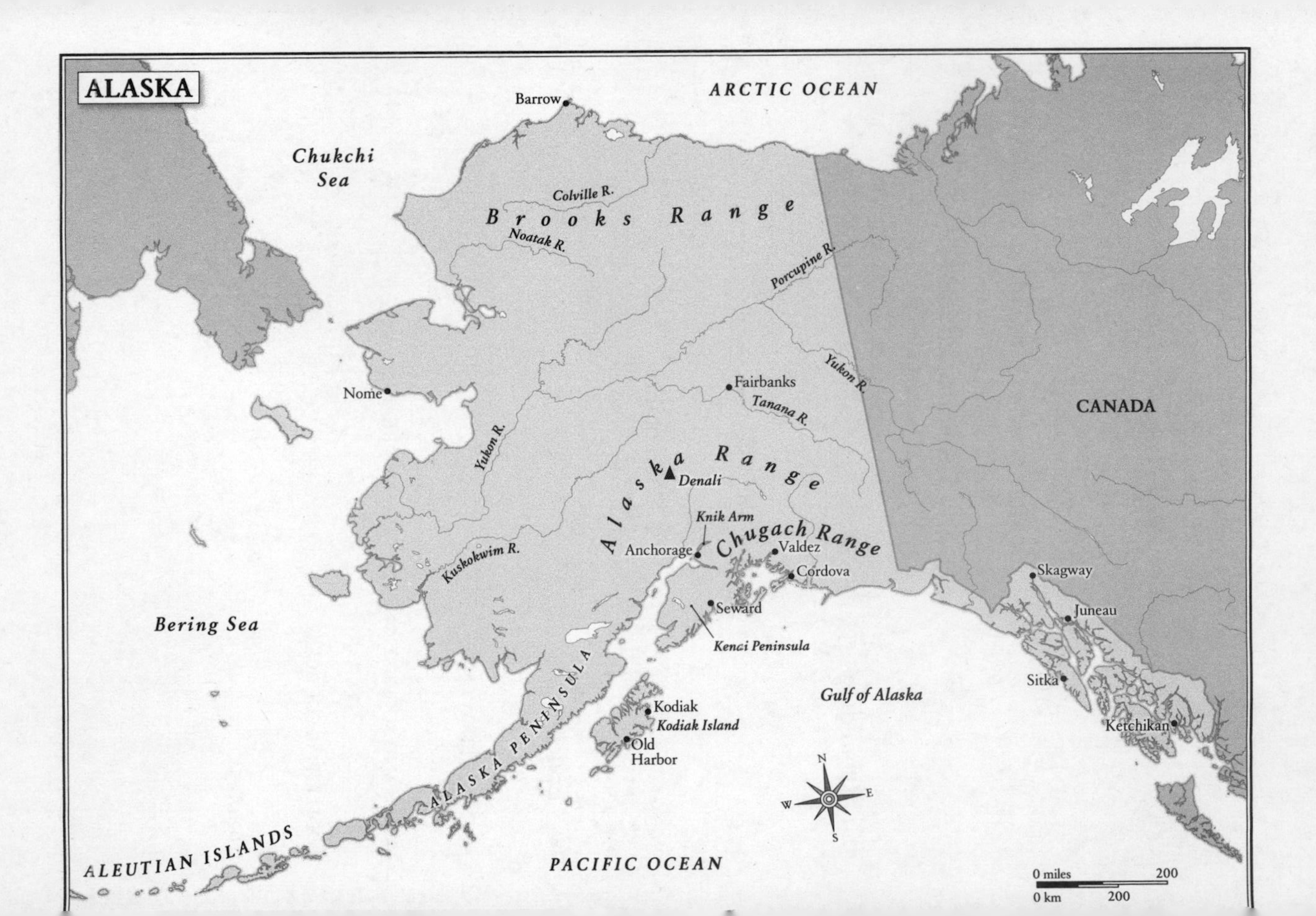
ALASKA
ARCTIC OCEAN
Barrow
Chukchi Sea
Colville R.
Brooks Range
Noatak R.
Porcupine R.
Nome
Yukon R.
Fairbanks
Tanana R.
CANADA
Alaska Range
Denali
Knik Arm
Chugach Range
Anchorage
Valdez
Cordova
Kuskokwim R.
Skagway
Seward
Juneau
Bering Sea
Kenai Peninsula
Sitka
Gulf of Alaska
Kodiak
Kodiak Island
Ketchikan
Old Harbor
ALASKA PENINSULA
N
W
E
S
ALEUTIAN ISLANDS
PACIFIC OCEAN
0 miles
200
0 km
200

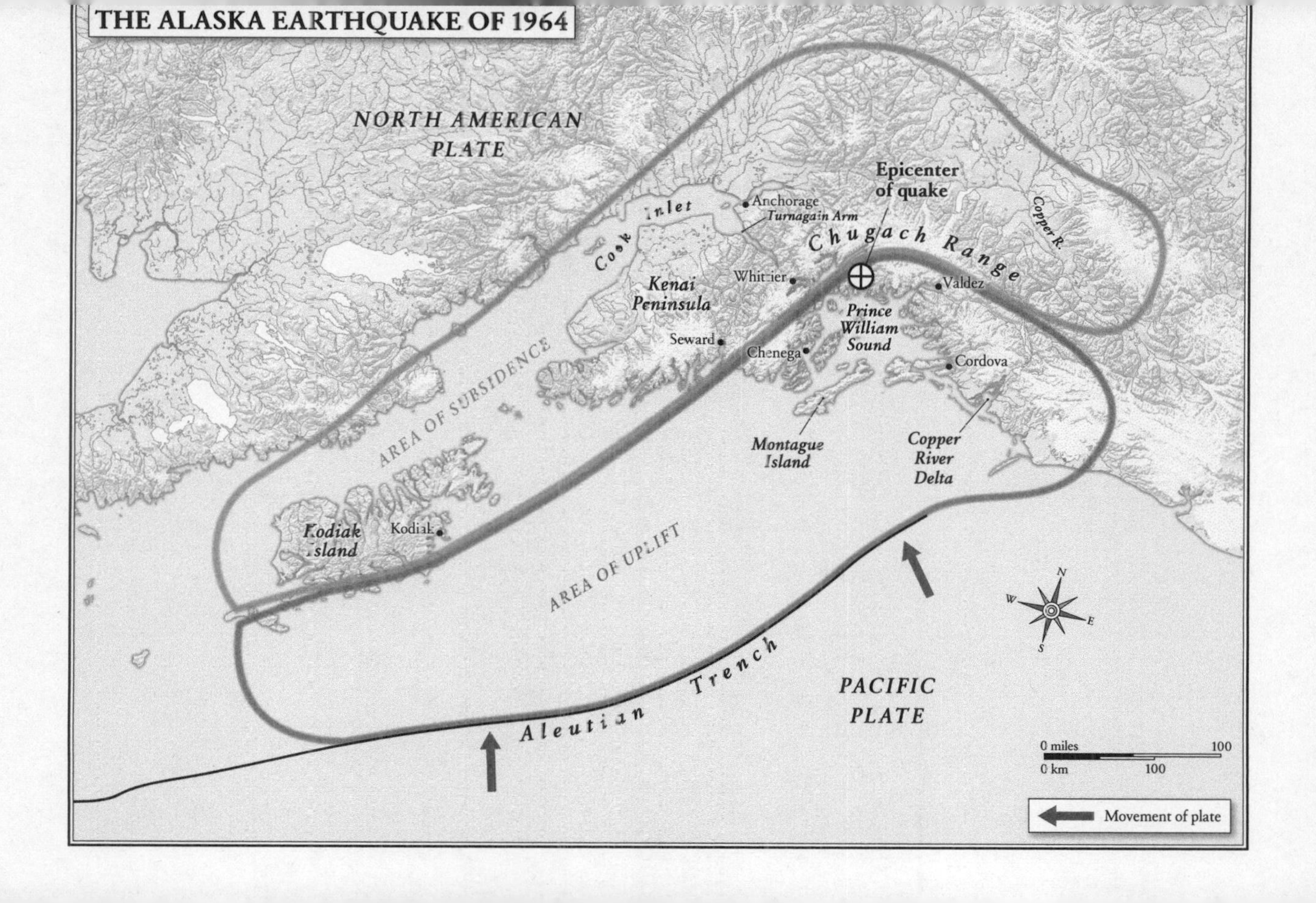
THE ALASKA EARTHQUAKE OF 1964
NORTH AMERICAN PLATE
Epicenter of quake
Anchorage
Turnagain Arm
Cook Inlet
Chugach Range
Copper R.
Kenai Peninsula
Whittier
Valdez
Prince William Sound
Seward
Chenega
Cordova
AREA OF SUBSIDENCE
Montague Island
Copper River Delta
Kodiak Island
Kodiak
AREA OF UPLIFT
N
W
E
S
Aleutian Trench
PACIFIC PLATE
0 miles
100
0 km
100
Movement of plate

1

ALTERED STATE

Riding shotgun beneath the clouds in a rattling De Havilland Otter, George Plafker gazed down upon an Alaska he'd never seen before.

A geologist with the US Geological Survey, at the age of thirty-five Plafker was already something of an old Alaska hand. Though he was based at the Survey's offices in Menlo Park, California, south of San Francisco (and with his wife, Ruth, had a modest house nearby where they were raising their three children), as a field geologist with its Alaska branch Plafker had spent many summers in the forty-ninth state. He was accustomed to exploring the backcountry for days at a time with little more than a rock hammer and a field notebook, some C rations to stave off hunger and a gun to ward off bears, studying and mapping rock formations to better understand, describe and catalog the state's immense mineral resources. To a degree Plafker even looked the part of an Alaskan sourdough, lean and solid with a shock of wavy black hair swept behind half-moon ears, brown eyes and a large nose above a nothing-fancy mustache. His huge hands looked as if they'd be more at home holding a lumberjack's ax or prospector's shovel than a compass and hand level.

In his time in Alaska, Plafker had come to realize he didn't much care for the vast tundra of the central and northern parts of the state. Much of this land was what the Russians had named *taiga:*

the boreal forest, thick with conifers and willows and birches and, to his mind at least, essentially impenetrable. Even if you could somehow get around the terrain, interior Alaska was boring, geologically speaking. You could search across an entire quadrangle—about fifty square miles—and never find a rock outcropping, he said. To Plafker, that was a colossal waste of time: outcroppings were a geologist's bread and butter, the key to understanding what the land was made of.

Southern Alaska—the grand arc of land from the Alaska Peninsula in the southwest, up through Cook Inlet and Anchorage and southeast to the Panhandle, encompassing Kodiak Island in the Gulf of Alaska and the smaller islands of Prince William Sound—was more to his liking, and it was here he had done most of his work. The region was alive with rocks that a geologist, or anyone, could see. These were rocks that had been pressure-cooked for millions of years, shoved down, lifted up, ground and muddled and re-formed and folded over and under and this way and that. Some of them—the dark, slaty ones that were so jumbled as to lack much of what a geologist might consider character—Plafker and others affectionately referred to as "black crap." Together with other kinds of rock they formed the region's signature feature—its steep-sided mountains that, where they met the sea, formed deep, narrow fjords. What's more, the mountains were draped by glaciers and laced with rivers, all of which wore at the rocks, grinding them into coarse gravel and fine silt and carrying it all down toward the sea in vast washes of sediment.

Bush pilots had flown Plafker across this geological wonderland too many times to count, dropping him off at some remote lake or beach or God-knows-where location with instructions to pick him up a week or so later. But in all of his time looking at southern Alaska from the air, he had never seen anything like this.

Plafker had arrived in Anchorage, the state's biggest city, from the Lower 48 the day before. In the late afternoon of the day before that—March 27, 1964, Good Friday on the Christian calendar—Alaska had been rocked by an enormous earthquake. No one knew precisely how strong it was. Of the state's two seismographs, one, in Sitka, had been disconnected at the time and the other, in Fairbanks, had gone haywire, the fury of the quake proving too much for it. But there were reports that the shaking had lasted the better part of five minutes, which is an eternity for an earthquake. In the great San Francisco quake of 1906—a defining catastrophe in the history of the United States—the ground had shaken for perhaps sixty seconds. The duration of shaking is an imprecise measure of a quake's power, but the longer the ground shakes, the greater the likelihood of widespread destruction.

Anchorage, from what Plafker had heard, had been hit hard. Whole blocks of the downtown were in shambles, with buildings and the streets they sat on torn apart by the tremors. One of the city's best residential neighborhoods was a jumble of tortured earth, toppled trees and splintered houses. The count of dead and injured in the city was unclear, but at first brush the numbers did not seem staggeringly high. Anchorage residents, though, had clearly been through quite a ride and had been terrified. Soldiers were patrolling a swath of downtown to keep the city from descending into panicked chaos.

But there were reports of greater death and destruction elsewhere. Much of this appeared to be due, not to the shaking per se, but to tidal waves that the quake had spawned. A coastal town in Northern California, two thousand miles away, had been hit by waves that had drowned an untold number of people. Closer to home, radio reports from Kodiak Island in the Gulf of Alaska suggested that more than one tidal wave had hit the island. The ports of Seward, on Resurrection Bay, and of Whittier, on Prince William Sound, were both said to have been severely damaged. But the most haunting news was coming from two other places in the sound: the port of Valdez, which sat in relative isolation at the head of a long fjord, and the small native village of Chenega, on an island of the same name.

The radioed reports about Valdez (pronounced val-DEEZ) seemed almost unbelievable: a long stretch of the waterfront, including docks, canneries and warehouses, had collapsed into the bay in an instant, taking more than two dozen residents with it. Valdez appeared to have suffered the largest loss of life of any Alaskan community—far greater even than Anchorage, which was many times its size. At Chenega (pronounced chuh-NEE-gah), a tidal wave had destroyed everything except the village schoolhouse on top of a hill. Along with their homes, a third of the villagers had been swept out to sea. There seemed to be little doubt that when the full toll of the disaster was known, Chenega would turn out to have the highest proportion of loss of life, by far, of any place in the state.

Plafker had come to Anchorage, along with two other Survey geologists, Arthur Grantz and Reuben Kachadoorian, to begin figuring out exactly what had happened. He'd had an inkling that things were going to be different when he'd flown in the day before on a Pacific Northern Airlines flight from Seattle. For one thing, the airline had announced that the plane would be landing at Elmendorf Air Force Base on the northeastern outskirts of Anchorage because the international airport southwest of downtown was closed. Its control tower had toppled in the quake, killing an operator unlucky enough to be working on Good Friday.

That was unusual enough. But what Plafker saw from the airliner as it approached Anchorage was truly otherworldly.

He had flown into Anchorage enough times to be familiar with the usual flight path. Approaching the city from the southeast, planes have to get over the Chugach Mountains, a string of peaks that arcs from southwest to southeast and serves as a kind of shield protecting interior Alaska. The easiest air route over the Chugach was at Portage Pass, forty miles southeast of Anchorage. Just south of the pass was the small port of Whittier.

Like much of the coastline in southern Alaska, the area around Whittier was often covered by a sheet of low clouds, as wind-driven air picked up moisture from the water that then piled up on the seaward side of the Chugach. Today was no exception. But looking out the window of the plane as it slowly descended toward Anchorage, Plafker was amazed to see a large, perfectly round hole in the clouds where Whittier should be. It was as if someone had taken a giant paper punch to the cloud layer.

It was only later, when Plafker saw the destruction of Whittier firsthand, that he put two and two together. Rising hot air, he realized, had created the hole in the clouds; when he'd flown over it, Whittier had been on fire.

Plafker and his colleagues landed at Elmendorf that afternoon, less than twenty-four hours after the earthquake. They were shown to officers' quarters—their civil service ratings made them colonels in the eyes of the military—and issued bunny boots and other cold-weather gear; although it was technically spring, southern Alaska was still blanketed in snow, and temperatures could easily hover at freezing or below. Then it was time to meet with the commanders of Elmendorf and the army base next door, Fort Richardson, to discuss the situation. The officers were happy to have the three scientists, even if they didn't know much about them.

When the quake occurred, Plafker and Grantz had been in Seattle, 1,500 miles to the southeast, at a two-day meeting of the Cordilleran Section of the Geological Society of America. Grantz was an older Alaska hand than Plafker. He had begun at the Survey in the late 1940s, a time when packhorses were still sometimes used for Alaskan fieldwork. He was at the Seattle conference to deliver a paper about his work dating rocks from the Chitina Valley, in the Copper River region southeast of Prince William Sound. Plafker, not scheduled to give a talk, had come to mingle and learn.

Late in the afternoon on Friday, the first day of the conference, word started spreading among the attendees about an earthquake in Alaska. The reports were sketchy, and there was little sense at that point of the scope of the disaster. But then a couple of scientists who had taken a break from the meeting to visit Seattle's biggest tourist attraction—the 605-foot-tall Space Needle, built for the 1962 world's fair—came back to the conference. From the observation deck 520 feet up, they reported, they had felt the tower sway. This must be one heck of a quake, Plafker thought.

That evening, back in his hotel room, Grantz got a call from the Alaska branch office in Menlo Park. George Gryc, the branch chief, was on the line suggesting that, given Grantz's and Plafker's knowledge of Alaska, they should immediately fly up to Anchorage. Kachadoorian, an engineering geologist with the branch who knew more about the impact of geology on structures, would fly up and join them. Grantz and Plafker had packed for only a two-day trip, so before going to the airport Kachadoorian would stop by their homes and pick up some fresh clothes from their wives.

Gryc was an Alaska veteran—he'd done some of the early geological mapping of the Brooks Range, in the Far North, during World War II—but he'd worked elsewhere within the Survey as well, including time at the agency's headquarters in Washington. He understood the Survey's strengths and weaknesses. He knew that the Alaska branch didn't have any real earthquake experts. But then again, neither did any of the Survey's other branches. Earthquake science, after all, was a young and small field. What the Alaska branch did have that was invaluable were people who knew how to get around the rugged state, who had gone up its creeks and walked its ridges, who wouldn't be spooked by the isolation of the backcountry or the prospect of running into a bear. Grantz and Plafker were two of them. Get on the next plane, he told them.

So here they were, listening to the military commanders' concerns. The bases had not been too badly damaged in the quake—although one of the barracks was now uninhabitable and the roof of a hangar had partially collapsed. But the officers were worried about the impact on other installations around the state and on the communications infrastructure that tied all of them together. In 1964, less than eighteen months removed from the Cuban Missile Crisis, the Cold War between the United States and the Soviet Union was as frigid as ever, and Alaska was the Western Front, just fifty-five miles from enemy territory across the Bering Strait. The military had spent billions of dollars on listening posts and radar systems designed to detect incoming Soviet bombers or ICBMs. Communications were largely through a series of relay towers that dotted the state, and the generals were worried. Clearly the violent shaking had caused large-scale land movement, as they could see in Anchorage, where in the residential neighborhood that had been destroyed the land had been thrust forward. Presumably this kind of movement had happened elsewhere, the commanders said, and while so far it seemed that communications were unaffected, they were worried that that might change. With further tremors or settling, the communications systems, which required precise alignments of the network of towers, might be in jeopardy.

There were other concerns too. The only road from Anchorage to Seward, the two-lane Seward Highway, had been extensively damaged. The port was now effectively cut off to vehicles from Anchorage and the rest of the state. The military needed to know what it would take to reopen the link. The Alaska Railroad, a lifeline for the fledgling state's economy that brought shipborne cargo from Whittier and Seward into the interior of Alaska, had been heavily damaged and was shut down as well.

The geologists had concerns of their own, most urgently about the potential for catastrophic flooding. There were many large rivers in southern Alaska, and if one or more of them were blocked by a landslide the result could be disastrous. A landslide would act like a natural dam, blocking the river and allowing water to build up behind it. But sooner or later the pressure of all the water would prove to be too much: the water would overtop and erode the blockage and would hurtle downstream in a raging flood that could wipe out any river communities. Plafker and the others had to find out if there were any major blockages and, if so, what could be done about them—divert the water, perhaps, or, in the worst-case scenario, move people out of harm's way.

The scientists' plan was to spend up to two weeks assessing the situation as best they could, with the goal of devising a full program of field research that could begin in the late spring and summer, when the ice and snow would be gone and surveying on the ground would be possible. The commanding officers asked how they could help. The geologists had a ready answer: aircraft and pilots, so they could do basic reconnaissance flights and land at affected areas, if possible, for a closer look.

That was why, on Sunday afternoon, two days after the quake, Plafker found himself in the Otter, a workhorse of a plane that, while grimy and noisy, was more than adequate for reconnaissance work. With him were Grantz and a pilot, an army lieutenant named Jones. Kachadoorian, with his interest in the effects of the quake on buildings and other structures, had gone off in a vehicle to get a close look at the damage in Anchorage.

The Otter had taken off from a small airfield at Fort Richardson and was soon over Cook Inlet, the large body of water that connects Anchorage with the Gulf of Alaska to the south. The pilot banked left and the plane headed southeast down Turnagain Arm, an inlet of the inlet, toward Whittier and Seward. Along the northern shore of the arm there was just enough flat terrain between the water and the steep slopes of the Chugach for the Seward Highway and the single Alaska Railroad track.

The road and rails were a mess. The first rockslide had occurred not far from Anchorage, and there were others along the route, as well as snow avalanches. In places debris had completely buried the road and the railroad tracks; in other places it had just pushed them toward the water and torn them up, so that the steel rails were bent and tossed about like so much spaghetti.

Plafker had a camera, a 35mm Olympus with some high-quality lenses, to document the destruction. Between shots he took notes on what he saw. Behind him, Grantz held an unwieldy sheaf of topographic maps that he used to track the plane's progress, marking the location of what Plafker was documenting on film.

It soon became apparent that more than rockslides had caused the damage. Both the road and the railroad crossed bridges over small rivers and streams on their way down Turnagain Arm, and Plafker noticed that something strange had happened to them.

Some of the bridges had toppled over, as might have been expected given all the shaking, but some of them hadn't. Instead, their decks had popped up. To Plafker they now looked, in a way, like dilapidated versions of those quaint arched footbridges that are common in gardens in Japan. Some of the toppled bridges, he realized as he looked more closely, had popped up before they'd fallen over. To him it appeared that the land had shaken so much that the riverbanks had, in effect, turned to mush. The banks had moved, sliding toward the rivers and taking the bridge pilings with them. As the pilings had gotten closer to one another, the bridge decks, which were still connected to the pilings, had been squeezed. They had nowhere to go but up.

As the scientists approached Portage, a small town at the head of the arm where the road to Whittier splits off to the east, they saw groves of trees along the shore that were now standing in seawater partway up their trunks. Before the quake they had to have been high and dry. That could mean one of two things: one, that the tides since the earthquake were now abnormally high, or two, that the land was now permanently lower, that it had sunk during the quake. The latter explanation seemed more likely; it also meant that Portage's homes and businesses were going to be permanently inundated as well. The town was doomed.

Soon they were out over Prince William Sound itself, with its spectacularly rugged scenery of glaciers, fjords, islands, bays and channels. The famous naturalist John Muir had described the sound, on a visit in 1899, as "one of the richest, most glorious mountain landscapes I ever beheld—peak over peak dipping deep in the sky, a thousand of them, icy and shining, rising higher, higher, beyond and yet beyond one another, burning bright in the afternoon light, purple cloud-bars above them, purple shadows in the hollows, and great breadths of sun-spangled, ice-dotted waters in front."

But now there were obvious signs that Muir's glorious landscape had been scarred by the earthquake. From the plane, Plafker and Grantz spotted trees and other debris in some of the sound's many bays and inlets, and evidence that parts of the coast had been hit by large waves or high water. There were areas along the shoreline, sometimes high up, where incoming water had washed the snow away. This "snow line" proved to be a convenient telltale for how high and far inland the water had come. At least the snow was good for something, Plafker thought. But high up along one inlet, Blackstone Bay, the hillside appeared to have been scoured; all the trees and other vegetation had been removed, leaving bare ground behind.

In his many flights into the backcountry over the years, Plafker had occasionally seen signs of a recent rockslide or snow avalanche. But now, for as far as he could see around the sound, the landscape was full of them. The earthquake must have caused thousands of slides—some little, some big, some that left piles of rock debris or snow and ice at the base of a slope, some that created huge swaths of destruction as the debris traveled quickly over a wide area. And perhaps some of the slides weren't really slides at all. The shaking had been so great that in a few cases it almost seemed that huge blocks of snow and ice had been flung off mountaintops, landing in the valleys below.

The region's many lakes, which had all been frozen over, showed the impact of the earthquake as well. On some, the frozen surface now looked like a jigsaw puzzle, the ice sheet fractured into hundreds of small pieces. But the ice on other lakes had remained in one piece, with raised ridges at the shoreline, suggesting that the ice sheet as a whole had moved back and forth during the quake.

Plafker knew he was witnessing destruction on a scale seldom seen anywhere. Clearly it was only because of the fact that Alaska was largely unsettled and empty that the toll in lives and property appeared to be relatively low; if a similar quake had happened in a heavily populated region the scale of the human disaster would have been overwhelming.

He couldn't help but be in awe of the energy that had been unleashed in just a few minutes two days before. But it was exhilarating to see this altered landscape up close. Over the droning of the engine, he and Grantz kept shouting at the pilot—to make another pass to get a better look at something, or to circle around while Plafker changed film. They didn't want to miss a thing.

They flew on, at times barely above the treetops, eventually turning toward the southwest and the Kenai Peninsula. From there they headed back toward Anchorage, where they landed about four hours after they'd taken off, exhausted but amazed at what they'd seen.

That first flight was followed over the next few days by others. Plafker usually sat in the copilot's seat so he could take photographs. Grantz juggled the maps from a seat just behind. For their part, the military pilots liked the work—all the low-level flying and detours to look at specific signs of quake damage were a welcome change from their usual tasks of ferrying equipment or military brass around.

Although he was not an earthquake expert, Plafker understood enough about quakes to know that what causes them is slippage along a fracture, or fault, in the rocks. Geologists see signs of faulting in rocks all the time, and Plafker had seen countless old small faults over the years in his fieldwork. This quake was so enormous and the effects were so widespread—they'd already flown across thousands of square miles of devastation—that the fault that had caused this one must be huge: so huge, in fact, that even if much of the rupture had happened out of sight (early guesses were that the slippage had occurred more than ten miles underground) there almost certainly had to be evidence of it at the surface. There had to be some disruption of the landscape along a more or less straight line, perhaps for dozens of miles, showing how the earth had moved this way and that. Yet as they flew around southern Alaska they saw nothing of the sort.

Plafker was intrigued—they had seen so much destruction wrought by the quake but no indication of what might have caused it. It began to gnaw at him a little. There was something different about this earthquake, he realized.

He couldn't have known it at the time—he was just a field

geologist, after all—but he'd be thinking about what made this earthquake different for the next few years. And for the rest of his career he'd be thinking about other quakes that were like it. The study of earthquakes, it would turn out, would become his life.

But first he and the others had to record what they'd learned about the Alaska quake from their two weeks in the state. Plafker, Grantz and Kachadoorian returned to Menlo Park to write up their findings, with Grantz taking charge of putting together a report, a "circular" in the parlance of the Geological Survey. And they made plans to return to Alaska with many more scientists in a few months, to further explore places like Chenega and Valdez and get a better understanding of why Alaska had been shaken to its core.

2

UNDER THE MOUNTAIN

Kristine Madsen looked up with trepidation at the white building with the red roof that would be her home for the next year. She'd wanted to teach in a one-room schoolhouse, and there it was, perched by itself at the top of a small hill. Things had moved quickly. One day she'd been in Anchorage, talking to an administrator with the state education department, and a few days later she'd arrived at Chenega, a village of about seventy-five native Alaskans, mostly Alutiiq, on a small island of the same name in Prince William Sound. It was late summer 1963, and now, after the floatplane that had brought her and her puppy, Tlo, from Cordova had dropped her off and taxied out on the water for takeoff, she wasn't quite sure what she had gotten herself into.

The island was isolated. Other than the occasional fish camp or cannery, the nearest place of any size was Whittier, more than fifty miles to the northwest by water, following a zigzag course through some of the sound's many inlets and passages. Cordova was nearly twice that distance, to the east, a more direct journey over the open water of the sound. Anchorage was inland to the northwest, on the other side of the Chugach Mountains. It might as well have been a continent away.

But Chenega's was a beautiful isolation. Steep hills, rising as high as two thousand feet above the water, were the dominant feature of its twenty square miles. They were studded with spruce and hemlock, the trees' heavy boughs shading a soft understory of ferns, mosses and berry bushes. The interior of the island was a mix of woodlands and the boggy peatland known as muskeg, with a few ponds scattered here and there. Small streams connected the wetlands and rushed down the hills to the sound. On clear days the views from Chenega were extraordinary, with the snowy peaks of other islands and the Kenai Peninsula visible in the distance to the southwest. But the cloudy, foggy days—common weather for Prince William Sound—served up their own kind of beauty, turning a cluster of small islands offshore into dark smudges, like daubs of dark gray in a wet-on-wet watercolor.

The land offered salmonberries, blueberries, currants and other wild delicacies for the taking, and dozens of kinds of animals roamed the hillsides, among them bear, deer, wild goat, mink, fox, porcupine, marmot and muskrat. In the nearby channels, whales and porpoises sometimes cruised by; seals, sea otters and sea lions swam or basked on rocks exposed by the tide. Gulls, sandpipers and other shorebirds skidded along the water's edge, while on land Steller's jays of brilliant blue cackled in the trees and bald eagles kept a lookout for prey from high above. There were clams, mussels and other shellfish in the intertidal zone, and fish of all kinds—halibut, cod, red snapper, hooligan and herring, and, above all, five species of salmon that arrived throughout the season. Herring roe, which stuck to blades of kelp like candy, was a delicacy all its own.

Like so much of the land around Prince William Sound, the steep hills of Chenega Island ran almost straight to the water, often with little more than a small strip of wild grass, or perhaps a few alders, as a buffer. From the air at first it seemed that the island must be uninhabited—where would someone live in a rugged spot like this?—but as Madsen's plane flew down Knight Island Passage and approached the island's south coast, a village came into view.

Chenega sat on a small cove, with quiet water that sometimes collected floating chunks of ice that calved from glaciers into nearby Icy Bay. A crescent beach several hundred yards long ran beneath a protective bulkhead, about eight feet high and made of stacked timbers. From the top of the bulkhead a long wooden dock led out into the cove. The dock was just a flat open deck, wide enough so that people could get by even if there was a pile of supplies on it. Toward the seaward end of the dock stood its only distinguishing feature, an outhouse.

Steps led from the beach to the top of the bulkhead, beyond which there was just enough flat land for a cluster of about half a dozen small wooden houses, with roughly the same number spread out along the top of the bulkhead to the southeast. The houses were steep-roofed and made of logs and planks, and most were simple, two-room affairs. Some had other attached structures: a smokehouse for preserving salmon and seal meat, and a bathhouse for the steam baths that were a ritual of village life. A small stream, cascading down from a dammed-up pond far up on a nearby hill, provided water for the baths and everything else.

Wooden boardwalks snaked around the houses and to the building that was, literally and figuratively, the center of Chenega. This was St. Mary's, a Russian Orthodox church. It was a modest wooden structure, long and narrow, with a dome-shaped cupola topped by the traditional Orthodox cross with three horizontal crossbeams. Inside there were no pews, but in the Russian Orthodox tradition the walls were covered with icons, and, for holidays, crepe-paper flowers made by the village women.

Near the foot of the dock was a building that was divided in two. On one side was a small cooperative store—it had lost half its sign in a storm and for years said only RATIVE STORE—that also functioned as a post office. The other was a boathouse, where in the past villagers had repaired their kayaks; now they used it to repair their skiffs. Nearby was another structure, a shed known to everyone as the Smokehouse, which had a pool table where the men of Chenega would spend time. The charge was ten cents a game, with the proceeds going to the church.

The store, post office and Smokehouse sat at the foot of a small hill, on top of which sat the schoolhouse. At an elevation of seventy feet, it was the highest building in Chenega, and the only one not near the water.

The building, built several decades before by the federal Bureau of Indian Affairs, indeed had one classroom, to the right of a small entryway, plus a narrow storeroom for books, supplies and government surplus food. To the left of the entry were living quarters, cozy and comfortable, with a bedroom, a living room and a kitchen. Behind the school, next to a small field that was used for recess, was a concrete-block structure that housed an oil-fired generator for the school. (A small hydroelectric plant built about two decades before by the Civilian Conservation Corps and meant to bring electricity to the whole village had been a failure, with the generator burning out after only a few years' service.)

Supplies, food, fuel oil—everything that was needed at the schoolhouse—were winched up on a wheeled dolly on a set of rails that ran up the hill from near the store. Students and others, however, reached the schoolhouse by way of a series of wooden steps. There were precisely ninety of them. Madsen would come to know the number well, because she counted them every time she went up.

Kris Madsen had grown up in Long Beach, south of Los Angeles, where her father worked at the naval shipyard near San Pedro. A tall, skinny girl with blue eyes and unbelievably red hair, she developed an independent streak from an early age. Perhaps it was because of her family situation: her birth mother had died when she was two, and her father, overwhelmed by the prospect of raising a toddler on his own, had sent her to live with his sister near Sacramento. When Kris was older her father got married again, to a schoolteacher named Bernice, and Kris came back to live with them.

Whatever the reason for her independent ways, by the time of her high school graduation, in 1958, Madsen was ready to get out and see the world. Her first stop was the University of California at Santa Barbara, just 120 miles up the coast. But for her it was the start of a new life.

At Santa Barbara, she gravitated toward anthropology and education. While she enjoyed studying the former, she soon realized that when it came to satisfying her wanderlust, education was probably a better choice. With a teaching degree, she could go just about anywhere. Everybody needed teachers.

When Madsen graduated after four years, she had no desire to begin her career in the local schools back home. She didn't even want to remain in California, but she didn't really have a clue where she might want to go. She'd only been as far as Vancouver, British Columbia, with her family. With no experience to go by, using the alphabet seemed as good a way as any to choose a place to work and live. She'd start at the beginning: Australia or Alaska.

The career office at the university didn't know much about jobs Down Under, but the Last Frontier was a different story. It so happened, a counselor told Madsen, that Alaskan officials were currently in California interviewing for teaching jobs. A state for just three years, Alaska was growing and in desperate need of teachers, especially those willing to work in native villages. Before, finding teachers for the native population had been the federal government's responsibility, through the Bureau of Indian Affairs, but now it was Alaska's problem. As an inducement, the state was offering excellent pay. Madsen could make about $5,000 a year.

She went to an interview and was hired on the spot. There was an opening in a three-teacher school in Old Harbor, an Alutiiq village on the big island of Kodiak in the Gulf of Alaska. Late in the summer of 1962, accompanied by her stepmother, Madsen made the long trip north—flying from Los Angeles to Seattle to Anchorage. She and Bernice did some shopping, picking up, among other things, a rabbitskin coat for the cold weather. Bernice said her goodbyes and left for California, while Madsen boarded a small plane for the 250-mile flight to the city of Kodiak on the island. She then took a floatplane the final forty miles to Old Harbor, on the island's southern coast, to begin her Alaskan teaching adventure.

As native villages go, Old Harbor was big, with a population of more than two hundred. Madsen taught the first and second grades and had more than a dozen students. Her first day at the school—her first day as a professional teacher—began with an incident that showed just how different life in Alaska could be. As her pupils filed into the classroom, a first grader named Rocky Christiansen came up to her desk. He was carrying a freshly caught salmon that was just about as big as he was, holding it up the way a professional fisherman would, with his fingers in its gills. Madsen had spent the morning neatly organizing her desk—she had wanted to make a good impression on any parents who might be bringing their children to school on the first day. With a flick of his forearm little Rocky quickly put an end to those plans, flopping the salmon onto the desk. "For you, teacher!" he exclaimed.

The school year ended in May, and for Madsen it was a good experience. She liked the other teachers, a husband-and-wife team, and befriended a Baptist missionary, a woman with whom she often shared meals. But her living quarters were beyond rustic—she slept on a bed behind some shelves in her classroom, which itself was little more than a glorified storage closet. Back home in Long Beach in the summer of 1963, working as a cook on a tuna boat, she decided she wanted to go back to Alaska for at least another year. But she thought she might want a different challenge.

So upon returning to Anchorage in late August she met with the state education official. She was willing to go back to Old Harbor, she told him, but she was also interested in the idea of teaching all elementary grades in one classroom. No sooner had the words come out of her mouth than the telephone rang. The official chatted with someone on the other end for a few minutes and then hung up. He turned to Madsen. We've just had an opening for a teacher at a one-room schoolhouse, he told her. Are you interested? It's in a small village in Prince William Sound.

When Europeans reached Prince William Sound in the nineteenth century, the natives hardly greeted them with open arms. Vitus Bering, a Danish captain who was commanding a Russian naval expedition and whose name now graces the strait that separates Russia and the United States, discovered Alaska in 1741 but missed the sound; the closest he came was Kayak Island, fifty miles to the southeast. It was thirty-seven more years before an explorer sailed into the sound. But what an explorer he was: the British Royal Navy captain James Cook, who did more than any other navigator to fill in the many blanks in the map of the Pacific Ocean. In May 1778 Cook was on his third and final voyage, an ultimately fruitless quest to find the Northwest Passage across North America, when he arrived off Hinchinbrook Island, at the southeastern edge of the sound. One of his two ships, the *Resolution,* was in need of repairs, so Cook looked for a quiet bay in which to anchor. He proceeded into the sound, finally stopping at a place on the eastern edge of it, an inlet that he called Snug Corner Cove.

In his journals Cook described encountering natives, about twenty of them, who approached his ships in two skinboats. "They were unwilling, however, to venture along-side," he wrote, "but kept at a little distance, shouting aloud, and clasping and extending their arms alternately." One of the natives then stood up in his boat, naked as the day he was born, with his arms out like a cross, and remained motionless for a quarter hour. Cook and his men took this and other behaviors as signs of friendship and returned them with "most expressive gestures," but the natives refused to come any closer. By the time Cook weighed anchor the next day, however, the natives had become more than just standoffish: with most of Cook's men occupied with repairs to the *Resolution,* at some point the natives attempted, unsuccessfully, to plunder his other ship, the *Discovery,* and steal some of the expedition's small boats.

Less than a decade later, when Russian sailors tried to barter for luxurious pelts of sea otter and other marine mammals, the natives of Prince William Sound largely spurned them, in some cases running their ships off. When they finally did trade with the Russians, the natives had little use for the items they received—they thought hardtack biscuits, for instance, were wood chips, and tossed them away in disgust. They viewed the Russian interlopers as aliens who had hooved legs (their boots), bandaged heads (their caps) and suckers on their bodies (the buttons on their jackets). Worst of all was what the natives made of their visitors' tobacco habit: these strange creatures breathed smoke.

The Russians and Britons—soon joined by Spaniards and Frenchmen in a free-for-all of exploration and exploitation—had come upon the Chugach, a subgroup of the Alutiiq natives who populated southern coastal Alaska from the eastern edge of Prince William Sound across to the Kenai Peninsula, Kodiak and neighboring islands and the Alaska Peninsula. The Chugach, of whom there were likely only a thousand or so at the time of European contact, were considered more aggressive than Alutiiq people elsewhere. One Russian captain wrote that this was probably because they were subject to frequent raids by other nearby groups, including the Eyak of the Copper River area and the Tlingit, to the southeast, along what is now the Alaskan Panhandle.

If the Chugach were fiercely protective of their territory they could be forgiven, because that territory had sustained them for thousands of years. Anthropologists are not sure precisely how long: the ice sheets that once covered the region retreated about 8,000 years ago, but the earliest archaeological evidence of human settlement discovered so far dates back only about 4,500 years, and the most elaborate diggings, near Hinchinbrook, date to about 500 BC. Whatever the archaeological record, it's fair to say that when Captain Cook arrived the Chugach were doing what they had been doing for generations: living off the bounty of the land and, especially, the sea.

The Chugach themselves divided the sound into eight territories, each centered on a village and politically independent. The westernmost one, bordered by Knight Island and Montague Strait, was called Tyanirmiut. And at least since the eighteenth century, the territory's main village was what other Chugach called Ingim-atya, meaning "under the mountain." Locals referred to it as Caniqaq, or "along the side" in Sugcestun, the regional dialect of the Alutiiq language. In English it was called Chenega.

Kaj Birket-Smith, a Danish ethnographer and anthropologist, visited Chenega and other parts of the sound in 1933 on an expedition to learn all he could about what were then known as the Chugach Eskimo. He talked to a couple of old-timers, including an octogenarian named Makari Chimovitski, who was thought to be the oldest native in the region. Makari, who spoke mostly Sugcestun, though he also knew a little Russian, told Birket-Smith of activities and practices that seemed to have gone on unchanged for centuries.

"The whole existence of the Chugach was based upon hunting and fishing," Birket-Smith wrote. And among the Chugach, the people of Chenega were considered more tied to the sea than others. There was a saying that the Chenegans were darker-skinned because they ate more marine mammals and fewer land animals and thus were "soaked in grease."

The subsistence life in Prince William Sound moved to the rhythm of the seasons. Halibut fishing began in late winter, and seal hunting started in earnest in May. Salmon, a main part of the natives' diet, began running in the early summer: Chinook, followed like clockwork by sockeye, chum, pink and, in August, coho. The herring fishery lasted late into the fall, when hunting for sea lions began. Sea otters, with their fabulous pelts, were hunted nearly year-round, and shellfish was harvested most of the year as well, especially when bad weather kept the natives out of deeper waters. Mountain goat was the chief prey of land hunters.

For sea hunting and fishing, as well as for general travel, the Chugach relied on *baidarkas*, or kayaks, fashioned of a light spruce frame, up to twenty-one feet long, which were covered in sealskins. (A small kayak might require the skins of six spotted seals, which were prepared by being buried for a time and allowed to ferment, making it easier to scrape off the fur.) For fishing, a one-hole *baidarka* would usually do, but for hunting larger sea mammals, two-hole *baidarka*s, and, later, three-hole versions, were the norm.

The biggest hunts of all, especially for Chenegans, were for whale. These usually took place in the winter and involved teams of four or six hunters in two or three *baidarka*s. As Birket-Smith described it, the hunt always took place in a bay or other quiet water, not in the open sea. A *baidarka* would quietly approach a whale, and the hunter in front would spear it with a lance, which consisted of a slate blade attached to a pole of spruce or hemlock. The blade would break off and, by Birket-Smith's account, all the kayakers would furiously paddle to the mouth of the bay and spread a poison—made, it was said, from the fat of dead whale hunters—on the surface that would prevent the whale from reaching open water. Other accounts suggest a more likely method of whale killing: the blade itself was dipped in a toxin obtained from monkshood or other plants. Either way, the injured beast might survive for a couple of days but eventually would die and drift ashore.

Birket-Smith was describing what in the 1930s were already considered the "old" ways; when he visited Chenega, he wasn't sure if whales were still being hunted there or anywhere else in the sound. The region and its people had changed much by then, largely because of the influence of, and eventual domination by, outsiders. Those early-eighteenth-century Russians who had been rebuffed by the natives of Prince William Sound had soon been followed by others who wouldn't take no for an answer. In 1793, just fifteen years after Cook, Russia established a fort on Hinchinbrook Island, putting down the locals in the process. Now the Russian American Company, given exclusive rights to the territory by the imperial government back home, had free rein to obtain the pelts of sea otters and other animals, which fetched high prices in China and elsewhere. But they needed the local population's assistance, in the form of their knowledge and hunting skills. The result, at least at first, was the virtual enslavement of natives throughout many parts of southern Alaska. In some villages—including, perhaps, Chenega—Russian traders and trappers known as *promyshlenniki* forced all the able-bodied men to leave their homes and work for them, and ensured their cooperation by holding women and younger men hostage.

Over the ensuing decades of Russian rule in Alaska, treatment of the natives improved, although native men were still required to work for the Russian American Company for a defined period of time, usually a couple of years. But in 1867, when the US secretary of state, William Henry Seward, negotiated the treaty that resulted in the purchase of the Alaska territory for $7.2 million, Russia's economic influence quickly waned.

American rule brought changes. Natives could now sell their pelts to one of several trading companies competing in a region where the Russian American Company had long had a monopoly. At first this was good for the Chugach, as prices rose. But sea otters, in particular, were quickly played out, and eventually the government declared a moratorium on hunting them. By the beginning of the twentieth century, with fur trading in decline, commercial fishing began to overtake it. This had an even greater impact on the natives' subsistence way of life, as they were no longer fishing just for themselves and their community. Almost overnight, canneries sprang up around the sound, most of them owned by a packers' association based in San Francisco. Salmon was the primary catch initially, and at first Chugach natives just sold their surplus fish to the canneries. But within a few decades most of the natives were working for the canneries for most of the summer. The men fished not in kayaks but in boats that were rented to them by the packers' group, often at exorbitant rates. They took their own fish only after they had satisfied the needs of the canneries. The women worked for the canneries too, preserving the fish.

For the people of Chenega, this meant that during the summer most families left to live and work at a cannery on a bay called Port Nellie Juan, which was reached by a circuitous water route via a narrow channel called Dangerous Passage. For more than three months every year, Chenega became a ghost town.

Not that the village had ever been a bustling metropolis. The first federal census, in 1880, showed a population of 80. By the turn of the twentieth century that number had increased to 141. But that was the peak; when Birket-Smith showed up in the 1930s, there were about 90 Chenegans.

Economic prospects, strong or weak, caused some of the population's rise and fall. The 141 people counted in the 1900 census included some non-natives who described themselves as prospectors—they were caught up in the gold fever that caused so many adventurers to come to Alaska.

Fevers of a different kind affected Chenega as well. Like so many indigenous communities around the world, those in Prince William Sound suffered from epidemics of diseases brought by outsiders. In the mid-nineteenth century, the Russians attempted, with some success, to vaccinate many natives against one of the most insidious of those plagues, smallpox. But the natives had no protection against other diseases. An outbreak of diphtheria and pneumonia in 1906–07 was especially brutal. People succumbed all around Prince William Sound, but Chenega was by far the hardest hit. Over a fortnight twenty-two villagers died, and many others who fell ill barely recovered. Few in the village were spared agony, and no one was spared grief. Chenega, as one Russian American publication described it, had become "the kingdom of death."

Half a century later, death would again stalk the little village in Prince William Sound. This time it would be violent, and would come from the sea.

By the time Kris Madsen arrived in Chenega in 1963, Alaska had been part of the United States for nearly a century—first as a territory and, since January 3, 1959, as the forty-ninth state. As if to reinforce that point, one of Madsen's duties was to raise the Stars and Stripes on a flagpole outside the schoolhouse each morning.

But decades-long Russian colonial rule in the nineteenth century had left indelible marks on the village, as it had on many other Alaskan native communities. One of the main influences was readily apparent in the list of students that Madsen received from state education officials. Of the fourteen children on her list, almost all had Russian-sounding surnames: Kompkoff, Eleshansky, Selanoff. Chenega was laced with Russian blood, the legacy of relations between the *promyshlenniki* and natives years before.

Everyone in the village knew one another, which was to be expected in a place with fewer than eighty people. People sometimes came and went—teenagers would move to Valdez or Cordova temporarily to attend high school, for instance, accompanied by a parent. But Chenega remained home.

The village was tight-knit in a more direct way: almost everyone was related. Grown children lived in homes that their parents had once occupied, while the parents now lived nearby (no one was more than a few doors away in Chenega). Families intermarried; there were cousins, aunts, uncles, grandmothers and grandfathers all calling one another neighbor.

Another lasting Russian influence was readily apparent: the church. The Russian Orthodox religion had been spread through Alaska by missionary priests beginning in the late eighteenth century, and, in part because the church worked to counteract some of the more brutal practices of the Russian traders, it stuck among natives. By one estimate, in the mid-1960s about a third of all Alaskan natives were church members.

Chenega had had a church for as long as anyone could remember, and in 1963 the small building was still very much the center of village life. For Chenegans returning from a hunting or fishing trip it was the first stop, to give thanks for their bounty. At weekly services—usually Saturday night and Sunday morning—the entire village was in attendance, men standing on the right and women on the left, listening to the words of a lay priest and singing Russian hymns. The services were where news of interest to the whole community was shared and where, once a year, a village chief was elected. In 1963, the chief was forty-year-old Charles Selanoff.

The most important person in the village wasn't the chief, however, but the longtime lay priest Steve Vlasoff. Vlasoff, who was born in 1888, spoke Alutiiq, English, Russian and Ukrainian. He had dark hair with a wisp of gray at the temples, a weathered, drawn face and an impish smile. He could be imposing in his black robes, with two crosses hanging around his neck, and the children of Chenega certainly found him so—for one thing, he required that all of them attend Sunday school. But Madsen found him to be charming. He'd come to her classroom and talk to the students about how to behave. "If you're not good, the teacher is going to pull your ears," he'd say. Then he'd turn his head to show his long droopy earlobes, which he claimed were a result of misbehaving during his religious training.

Even in the 1960s, the way of life in Chenega was still largely one of subsistence. Villagers earned money from fishing, which they did in boats leased from the cannery (and which bore the initials NJ, for Nellie Juan). The income enabled them to buy staples like flour and sugar from the store and other necessities through the mail (the Sears and Montgomery Ward catalogs were much-thumbed in the village), delivered by boat or floatplane. But Chenegans still took a lot of fish for themselves, in late summer after the canneries had been satisfied, and smoked and dried it to last all year. They relied on the land and sea for most of their other food: gathering berries and wild duck eggs, trapping small animals, fishing for halibut and other fish, digging up clams and hunting goats and deer, as well as seals and other marine creatures. Herring eggs on kelp, boiled up and served with some seal oil or butter, made an especially tasty dish.

Madsen soon settled into life on the island. Tlo, the half-collie, half-Lab puppy that she had gotten from a family she'd stayed with in Anchorage (and named using the initials of the head of the household), was constantly by her side. She'd become friends with an older man in the village, Norman Selanoff, who sometimes made duck soup for her. She'd also settled into the routine of teaching, although in a one-room schoolhouse not much was routine. She grew fond of her students—they were a playful group, honest and filled with energy, and the older students were good about putting up with, and helping, the younger ones. Timmy Selanoff, a son of the chief, was a favorite. The eleven-year-old had a bit of mischief in him, Madsen thought, but never got in too much trouble. None of the children did, really. But she found that students often brought trouble from home. Excessive drinking was a problem—it seemed to get worse around the time that the mail boat would arrive, Madsen thought—and sometimes when her students would show up in the morning it seemed as if they hadn't slept at all the night before because the adults in the household had been drinking.

But by and large Chenega was a happy place for a child. There was the beach to play on, with plenty of space to chase birds, play marbles or ball or look for shellfish when the tide was out. Children started fishing and hunting at an early age, and often whole families would go out together, taking turns with their .22-caliber rifles picking off seals. In the winter, the hills were fine for sledding, and the pond above the schoolhouse became an ice rink of sorts. No one had skates, but they'd play on the frozen pond anyway.

Russian Christmas, celebrated in early January according to the Julian calendar used by the Orthodox Church, was an especially joyful, even raucous, time. Everyone participated in the Orthodox tradition of "starring," which is derived from the Christmas story of the Star of Bethlehem. Villagers would make ornate stars and carry them around in groups, stopping at each house to spin the star, sing Christmas carols, eat pies made by the woman of the house and drink. Adults would throw coins on the floor, which children would scurry around and pick up. The festivities would continue for several nights.

At the school, Madsen had help from a man in the village, Joe Kompkoff, who served as janitor. He was known to almost everyone by his nickname, Sea Lion Murphy. He'd gotten the name, or so one story went, when in a bar in Cordova he'd bragged to a bunch of non-native Alaskans about the merits of roast sea lion—that it was the "beef of the sea." The others had been intrigued, so they had gotten a beef roast and some sea-lion meat and set up two slow-cookers on the bar. When they dug into the meat at the end of the day, they couldn't tell one roast from another. The other men were so impressed that they gave Joe Kompkoff the nickname.

Kompkoff, thirty-seven, lived in a house along the bulkhead with his young wife, Avis. She was not Alutiiq but Eyak, a group whose home territory was in and around Cordova, to the east. Avis had been born in Anchorage to a mother who was an alcoholic—she couldn't handle much of anything, least of all a baby—and had been adopted within a few days of her birth by a Chenega couple, Sally and William Evanoff. Now, not yet twenty, Avis already had two children of her own—Jo Ann, the oldest, was nearly three, Joey was not quite two—and was pregnant with a third, a boy they named Lloyd when he was born in the fall of 1963. She and Joe and the kids lived in a two-room house that her adoptive parents had once lived in. It had a living room–kitchen and a bedroom, with an old wood stove made from an oil drum.

Avis had loved growing up in Chenega—hunting with her father, gathering berries that her mother made into jam and pies and playing underneath the dock when the tide was out. As she grew older, she began to appreciate Chenega's beauty, and now, as an adult, from time to time she would stop what she was doing and gaze out past the cove to the lands beyond, including the peaks of the Chugach. *What a lovely world God has created*, she would find herself thinking. *And we get to enjoy it in this beautiful place.*

3

AN ACCIDENT OF GEOGRAPHY

The scene that greeted William R. Abercrombie upon his arrival at Valdez, Alaska, on the evening of April 21, 1899, was, as he later put it in his journal, one that he would not soon forget.

Abercrombie, a forty-two-year-old captain in the US Army, had visited the settlement the year before, as part of his explorations on behalf of the War Department, and at the time had encountered many would-be gold prospectors, fortune seekers who had been lured north by the promise of riches to be found in Alaska and the Canadian Yukon. The gold rush was in full swing, and these men had been lured to Valdez, specifically, by another promise, made by unscrupulous steamship operators on the West Coast. The little outpost at the head of a long fjord on Prince William Sound, the operators said, was the surest gateway to the gold.

The first of these adventurers had shown up late in 1897, and when Abercrombie had arrived that first time, early the following spring, most of them had already headed north. Others arrived while Abercrombie was there and left as soon as possible, hoping to travel while the ground was still frozen, which would make the going easier. Still others left in the fall. All of these adventurers pinned their hopes on successfully traversing the Valdez Glacier, twenty miles of ice that began four miles from where they had disembarked from their steamships and that rose to an elevation of about 4,500 feet. A second glacier, the Klutina, would have to be crossed as well, and then the would-be miners would have to build boats—having lugged saws and other tools with them for the task—for the trip down the rough rapids of the Klutina River to the Copper River valley. Some would stop and hunt for gold there, but others would keep going to the Tanana River and, eventually, to the Klondike.

That was the plan, anyway. Most of the men (they were almost all men) were novices, lawyers and teachers and accountants and businessmen with some money but not a lot of sense, it having been consumed by that most pernicious of maladies, gold fever. They had little idea what they were getting into.

Abercrombie himself was partly responsible for the popularity of this route. He had first come to the Alaska Territory in 1884, as a lieutenant heading up a small party to explore the Copper River. He had not gotten very far before quitting in disgust. The rapids proved to be too much, and the water was so cold his men could pull upstream for only fifteen to twenty minutes at a time. But on his return he reported that the Valdez Glacier might be a good alternative to reach the interior. It isn't clear that he actually saw the glacier, since the one he described ran east–west, whereas the Valdez Glacier ran north–south. Perhaps he saw no glacier at all. But the idea of using a thick ribbon of ice as a highway to the gold eventually caught on.

In 1898, with the War Department worried about the large numbers of hapless Americans heading north, Abercrombie came to Valdez under orders to seek an alternative to the glacier route. Before heading back to Seattle, he had found the remnants of a trail that skirted the glacier far to the southeast, and in the spring of '99 he returned aboard the steamship *Excelsior* to begin the work of building a road. Barely had the ship reached Valdez when scores of those once-proud adventurers he had seen the year before began clambering aboard. They had had enough of Alaska and were desperate to get home.

The men were demoralized and despondent, and by their looks

had been through a difficult time. Their woolen Mackinaw suits were worn out and faded from overexposure to the elements. They were unwashed, their hair and beards long and unkempt. Many of them had the atrociously bad breath that is a sign of scurvy, and frostbitten toes and fingers were common. "A more motley-looking crowd it would be hard to imagine," Abercrombie wrote.

The arrival of the steamship with a military contingent had given the men a new sense of hope—of getting home, preferably at the government's expense, since most were now almost penniless. Once away from the squalor of Valdez and on board the relative splendor of the *Excelsior* they couldn't contain themselves. "A wholesale orgy was inaugurated that lasted until midnight," Abercrombie wrote, "the cabin and decks of the steamer giving unmistakable evidence of the potent influence of the liquor on those who had indulged so freely."

Abercrombie soon learned what had happened, from speaking with a quartermaster's agent who had stayed behind the previous year. "My God, Captain, it has been clear Hell," the agent told him. The short story was that the men's dreams of gold had been dashed. of an estimated four thousand adventurers who landed in Valdez in 1897–98, fewer than one in ten had gotten anywhere near the Yukon or the Klondike. The longer, sadder story was that the route to the interior had proved deadly. Some men had died on the twenty-mile trek across the Valdez Glacier, lost forever in crevasses or frozen to death after being stalled by howling winds and storms. Others had made it over the glacier but found the going almost as bad on the other side, with many eventually dying of exposure or scurvy or drowning in the Klutina or the Tanana. The men now in Valdez—about six hundred all told—were the lucky ones. They had managed to straggle back.

Perhaps they weren't so lucky. Throughout the long and exceptionally hard winter of '98–'99 they had lived like rotting sardines, crammed fifteen or twenty into cabins so small there was barely room to move. At least 70 percent of the survivors, Abercrombie estimated, were "more or less mentally deranged." Many of them talked of a "glacial demon," a sort of Abominable Snowman who whisked lives away.

Abercrombie knew something of the horrors up on the ice. He had crossed the Valdez Glacier the summer before as part of his reconnoitering of the region, and though the conditions had not been as harsh as in other seasons, the trek had been nightmarish. With the help of some wild ponies, he and a small party of soldiers and civilians had made it across in twenty-nine hours. The physical dangers were great—at one point they found themselves in an ice field so littered with crevasses that it seemed equally dangerous to go forward or back. But the threats to the mind were great as well; during the night, no one slept a wink as they heard the constant cracking and booming of the ice and expected at any moment to be entombed by blocks falling from above them. "During my 22 years of service on the frontier," he wrote later, "I never experienced a more desolate and miserable night."

It was difficulties and dangers like this that led the War Department to send Abercrombie back in 1899 to begin building the road to the north. This route avoided the glacier by following the old trail through Keystone Canyon, a narrow defile a dozen miles east of Valdez, and then north across a low point in the Chugach Mountains, later to be named Thompson Pass. Abercrombie needed men to help, so without much pause he arranged passage back to Seattle for some of the worst-off of the survivors, started building a hospital to help others recover, and offered work to those who were healthy. It was the beginning of better times for the little settlement at the water's edge.

As in other parts of Prince William Sound, natives had frequented the Valdez area for untold centuries. The canyon trail that Abercrombie found, in fact, was a native one, used for trading and raiding. This was Alutiiq country, frequented by the same natives who inhabited Chenega and elsewhere in this part of the sound.

But little is known about what, if any, settlements were established by the natives. So the first permanent inhabitants may have been those would-be miners of the gold rush.

Europeans had discovered the area more than a century earlier, although Captain Cook had missed it when he entered the sound in 1778. It was left to Salvador Fidalgo, a Spanish navigator, to first sail into the fjord that is now called Valdez Arm, in 1790. Like Cook, Fidalgo had been looking for the Northwest Passage, and like Cook he never found it. But sailing eastward, about a dozen miles into the arm he passed through a narrows and found another long deep bay, which he named Port Valdés, after a Spanish admiral. The setting was spectacular, with nearly mile-high peaks on both sides and with glaciers seemingly poking out of every low point in the mountains. The bay itself was twelve miles long, three miles wide at its widest point and up to nine hundred feet deep. (In the twentieth century, the region's boosters pointed out that the bay was so big and so protected that it could easily provide mooring space for all the world's ships.) Best of all, at its eastern end was a wide alluvial plain, a delta made of sediments washed down from some of the glaciers. In a place where most of the coastline was inhospitable because the mountains plunged right to the water, this was flat and friendly terrain.

For more than a century after Fidalgo's visit, little of note happened in Port Valdés, although Russian traders no doubt circulated there to do business with natives. But in the mid-1880s gold was discovered in northern British Columbia and in southeastern Alaska. Word began to spread to Seattle and beyond, and when other discoveries were made farther north—especially a huge strike in 1896 on a tributary of the Klondike River in the Yukon—the gold rush was on.

The Klondike, however, was in Canada, and most of the major routes to it were largely in Canadian territory as well. Alarmed by the influx of novice miners who might strain the region's meager infrastructure and resources, the Canadian government imposed rules and customs duties for those who would enter from the United

States. Among other things, adventurers were required to pack in a year's supply of food. That added to the cost and also made the trip more difficult, as that much food could easily weigh half a ton or more.

The steamship companies of the Pacific Northwest began advertising the route over the glacier from Port Valdés as an "all-American" one that would allow miners to avoid the onerous Canadian rules and duties. What's more, if they so desired, prospectors could forget about going to the Klondike—where all the good claims were rumored to have been taken anyway—and could instead explore the valley of the Copper River, which was much closer and said to be rich in gold.

When the first ships arrived at Port Valdés during that fall of 1897, they discharged passengers and their supplies as close as possible to the foot of the glacier—a spot on the shore about four miles away. At least the start of the adventurers' journey would be a little less onerous. Thanks to the bay's high tides, there were large tidal flats there. Ships could beach themselves when the tide was going out, unload directly onto the land (or, in the winter, ice) and leave with the rising tide.

The would-be prospectors would still have to ferry their supplies, bit by bit, across the gently rising plain to the foot of the glacier. Tents and then ramshackle buildings sprouted near the shore, as staging all the equipment and supplies and getting them to the glacier could take days. Some people never even made it to the glacier but instead chose to stay behind, figuring they could make more money catering to the miners' needs than they ever could by heading north to the Klondike.

Thus the settlement of Valdés—soon changed to Valdez, amid anti-Spanish sentiments during the Spanish-American War—came into existence, the result of an accident of geography. It was the shortest distance to the glacier and, hopefully, the gold.

Later, though, it was geology, not geography, that was to play a crucial role in the town's fate.

A glacier such as the one above Valdez is like a giant milling machine moving across the landscape. The weight of the ice causes the glacier to move, very slowly, downhill with gravity, and as it does it scours the rock below and to the sides, grinding it up into smaller pieces of all sizes—boulders, gravels, sand and fine silt. Some of the boulders and finer material are carried along by the glacier and eventually deposited in a heap, called a moraine. But as the glacier creeps and grinds along, friction causes the underside of the glacier to melt too. Much gravel and sand, especially, is washed out by this meltwater and carried by gravity to the lands below.

In colder times, when the glaciers extended farther south—when much of Alaska was covered in ice—this scouring action had carved the deep narrow basin that became the Valdez Arm and Port Valdez, as well as the other fjords in Prince William Sound. Now, in warmer times, the glaciers had retreated, but they were still slowly scraping the Chugach Mountains. Some of the resulting silt, sand and gravel was brought down from the Valdez Glacier and others nearby by meltwater that formed three rivers—the Robe and Lowe to the east, and an unnamed meandering stream that came down from the north.

Over thousands of years, all that sediment had filled in part of the deep basin and formed the broad plain, which was roughly triangular in shape. One point of the triangle was at the foot of the Valdez Glacier; two long sides, defined by the steep slopes to the east and west, widened out from there. The third side of the triangle was at the water's edge, and it was about three miles long. The settlement of Valdez sat about midway along this side.

This triangular delta was deep, as well: in some places there were up to six hundred feet of gravel, silt and sand. Below that was bedrock. The sediments were loose, or what geologists call unconsolidated, meaning they hadn't been subjected to the pressure and heat that over eons would turn them into solid blocks of rock.

Unconsolidated sediments can support weight because the individual grains of sand or pieces of gravel are in contact with one another, forming an unbroken, weight-bearing chain from the surface all the way to bedrock. So in that respect, the plain was a fine place to start a town—it was easily solid enough to support the buildings and whatever else was constructed on it. But the individual grains in unconsolidated sediments also have plenty of space around them, and these pore spaces can become filled with water. That can spell trouble under certain conditions. If the sediments are shaken, the grains become compacted and the pressure of the water increases. The sediments slump—which can cause anything built on them to become unsteady—and water is forced out in any direction in which the sediments are not confined. Usually this is upward, but if there is an open face—a riverbank, say, or along the water's edge at Valdez—the water can move outward, taking sediment with it. What had seemed like perfectly solid ground can turn, in an instant, into quivering jelly and give way. The process is called liquefaction.

The Valdez plain had plenty of water. For one thing, the unnamed glacial stream meandered across it, changing course and splitting into smaller braid-like streams depending on the seasonal flow. This was recognized as a problem by an engineer in Abercrombie's party, who wrote of the hastily built structures at Valdez: "There is danger at any time of having the buildings swept into the Bay by the swift and quickly changing channels formed by the numerous streams." And it was a problem when Valdez grew—there was so much flooding, the location of which was unpredictable year to year, that during the Depression Civilian Conservation Corps crews had built dikes to protect the town.

But that was only one problem; belowground another danger lurked. The water table in most parts of the plain sat just a few feet below the surface. That meant that the sediments were not only loose but constantly saturated as well. The plain was ripe to liquefy if shaken. And because it was bound on two sides by the solid rocky slopes of the Chugach, if it were to liquefy, the water and sediments had only one way to move: toward the bay.

Although no one knew it, Valdez had been founded on land that was inherently unsafe. All it would take to trigger a catastrophe would be some rigorous ground shaking—something that nature would eventually provide.

The debacle of the winter of 1898–99 might have spelled the end of the tiny settlement, but the road that Abercrombie was building proved to be a lifesaver. In 1899 he completed some ninety-three miles of a rough-hewn trail, wide enough for packhorses, to the Tonsina River. In a few years there was a link all the way to Eagle City on the Yukon in the gold country. When the gold rush there ended, construction crews from the Alaska Road Commission (which had taken over the work from the army) extended the road to the growing town of Fairbanks, in north-central Alaska. By 1907 the road was finished, and Valdez had become the main port for goods destined for the interior. A steady stream of horse-drawn wagons headed out of Valdez through Keystone Canyon from late spring to early fall, and some horse-drawn sleds even made the trip when snows came.

The little town that had sprung up by accident on the alluvial plain grew fast; by 1905 the population exceeded five thousand. Although by then the Klondike gold rush was long since over—many of the original prospectors had given up or decamped to Nome, far to the north and west, when gold was found there in 1899—some gold discoveries in the Copper River valley and near Valdez itself kept the town a base for miners. Copper was being mined as well, and reports of a huge lode of rich copper ore in the Wrangell Mountains to the northeast suggested that Valdez might remain a mining hub for a long time.

In place of the tents and ramshackle cabins of the first few years, solid houses and commercial buildings were built along streets laid out in a grid. There was a thriving commercial center, with stores, hotels and at least half a dozen saloons. Down at the waterfront, a pier and dock were built to make it easier for ships to unload. The military, eyeing another stretch of relatively flat terrain across the Lowe River on the south side of the bay, started building an army base there, called Fort Liscum. From here the army's Signal Corps strung cables to the north, and under the water to the south, connecting Valdez by telegraph to the rest of the world.

Among the thousands of adventurers who were lured to Valdez to seek their fortune during those early days were two Nebraskans, George Cheever Hazelet and Andrew Jackson Meals. Hazelet was in his midthirties, a high school principal in the central part of the state, when news started filtering in of riches to be had in Alaska. His friend Jack Meals was nearly ten years older and was a little less of the settled type—he had no formal education and had worked as a rancher, farmer, wagon train scout and bronco rider.

They arrived by steamship in March 1898 and proceeded, like everyone else, to unload their equipment on the ice and haul it painstakingly to the foot of the glacier. But Hazelet and Meals turned out to have more perseverance than many of the other gold seekers. After much struggle they made it up over the glacier and into the Copper River valley. Rather than heading to the Tanana and the Yukon and the Klondike, however, the men stayed in the Copper River region. After a year they had found little gold but had proved to themselves that they could survive in rugged Alaska. Despite numerous hardships—at one point, low on food, they spent several fruitless days hunting moose and had to settle for a measly marmot (which, being from the Plains, they called a prairie dog)—they both weighed more than when they had started.

Hazelet and Meals traveled and worked all over the region, mining claims and helping to build roads and dams. Valdez became their home base, and they became pillars of the community. In 1901, the two men homesteaded a square mile of land about four miles northwest of the town site. The property consisted of gravels and sands but featured some exposed bedrock too. There was plenty of flat land for homes and businesses, and along the water there was room for ships, small and large. They named the area Port Valdez and had a grand scheme to develop it. Those plans never came to fruition, but eventually, more than half a century later, the land would prove to be invaluable.

Valdez kept booming well into the second decade of the twentieth century, with gold and copper mining being a main engine of growth. Gold mines, in particular, sprang up all over. In the Klondike most of the gold had been in the form of placer deposits—flakes and nuggets that over eons had been naturally eroded out of rock into gravel beds, which the miners sorted through. But around Valdez the norm was hard-rock mining, which required excavating ore and crushing it to extract the gold. Miners made almost superhuman efforts to haul rock-crushing machinery, known as stamp mills, into and over the mountains around Valdez. Some of the operations were fabulously successful, including the aptly named Cliff Mine, overlooking the bay about eight miles west of Valdez, which employed as many as sixty people and at its peak, from 1910 to 1915, produced a million dollars' worth of gold.

Valdez now had three newspapers, several breweries, a bowling alley, a hospital, a library and even a YMCA complete with baths, a gymnasium and a reading room. A university opened in 1916, though it didn't last long. The town became a district seat of the federal court system, which administered the territory's judicial affairs, and a lot of lawyers became at least part-time residents. With the advent of the world war, the population dropped off somewhat, as it did throughout Alaska, but there were still about 2,500 people in Valdez by the end of the decade.

The road north, then called the Richardson Trail, was the other source of the town's prosperity, as it was still the shortest route to Alaska's interior. Stage companies, using horse-drawn sleds or wagons depending on the season, provided regular service to Fairbanks, moving freight, mail and passengers across the 360-mile route in about a week and a half. Horses remained in use for many years, as motorized vehicles didn't begin to travel regularly on the road until the 1920s.

Amid the boom times, Valdez suffered setbacks, including the destruction of the business district in 1915 in a spectacular fire that may have been deliberately set. Despite the best efforts of the fire department and soldiers from Fort Liscum—they exploded sticks of dynamite in some buildings in a desperate attempt to beat back the flames—most of the structures could not be saved.

Valdez recovered quickly, rebuilding most of the downtown within months. In the ensuing years, however, other problems proved to be more long-lasting. The army closed down Fort Liscum in the mid-1920s, part of a wave of postwar cutbacks. Gold mining went through a gradual decline in the 1930s (and ended completely during World War II, when the federal War Production Board shut down gold mines, considering them inessential). Fox farming and fishing took up some but not all of the slack. Most important, the rise of the Alaska Railroad cut into Valdez's role as a hub for shipping to the interior.

Still, the military's need to build bases and airstrips during World War II and surveillance systems during the Cold War meant that for a while at least there was plenty of cargo passing through Valdez. At one point during the war, it was estimated that a loaded truck left the port every fifteen minutes, bound for the north. The army stationed a work battalion in town, boosting the population by about five hundred.

One of Valdez's selling points was that its harbor remained ice-free all year long—it was the northernmost port in North America that could make that claim—and after 1949 what was now called the Richardson Highway remained open to trucks through the winter too. A truck depot on the outskirts of town became the site of annual "roadeos," driving-skills competitions that brought truckers from all over the state. By the late 1950s and into the '60s, though, shipping had again fallen off. There was less business to go around, and Valdez found it hard to compete with the railroad.

By then Valdez had evolved into a smaller but in many ways calmer community, without too much going on but cozy and safe—the kind of place where neighbors knew one another, where front doors were never locked and people left their keys in their cars and where Saturday-night dances attracted just about everyone, including young people. High school students often brought their rifles to school, keeping them in their lockers until lunchtime, when they'd leave for an hour of ptarmigan hunting on the other side of the dike, bringing the rifles—and any dead birds—back to their lockers afterward. In the winter, one father would tie sleds to the back of his station wagon and drive his children around town, revving it up around corners to try to flip the kids off. It was all great fun.

The federal courthouse had burned down in 1940, and the district headquarters had been shifted to Anchorage, the biggest city in the territory, but the Road Commission was still a major employer in Valdez, as was a facility for the mentally disabled, the Harborview Nursing Home, that opened in 1961. There was still enough commercial fishing to support a cannery or two, although fishing went through some very lean years and even at its peak was never as big an industry in Valdez as it was in Cordova and some other communities. Valdez joined in the post-statehood push to develop tourism, billing itself as the "Switzerland of Alaska" because of its dramatic setting amid the steep peaks that were "almost more Alp-like than the Alps themselves," as one brochure put it. A tourist boat, the *Gypsy,* took up to fifty visitors at a time on daylong trips through the narrows to where one of Alaska's most magnificent glaciers, the Columbia, spilled into the water. There they could gaze in wonder at the front edge of the ice sheet, which was two miles wide and up to 650 feet thick. A blast from the ship's horn might even cause pieces of the ice to break off and fall spectacularly into the water.

Valdez was almost exclusively white, with only ten or fifteen natives among the population, according to one estimate. In a census taken in the late 1950s, the town was found to have 841 people, including 331 children.

Most of the population lived in a roughly thirty-block town center, with two perpendicular streets at its heart: Alaska Avenue, which connected to the start of the Richardson Highway on the east side of town and ran west to the water, and McKinley Street, which ran north–south and intersected Alaska a block from the waterfront. Homes were modest, mostly two-story affairs; many had been built by their owners. They had steep roofs, a necessary feature because Valdez was one of the snowiest places in Alaska, averaging about twenty-two feet a year. In 1928–29, townspeople suffered through a winter with more than forty-three feet of snow.

Valdez was more connected than ever to the rest of Alaska. The Richardson Highway was now a gateway not just to Fairbanks but to Anchorage, although it was a roundabout route—north for 120 miles along the Richardson to Glennallen, then west for another 140 miles on the Glenn Highway, built by the military during World War II. It took at least eight hours to drive the 260 miles, but since Valdez did not have much in the way of medical facilities—some years it was hard to keep even a general practitioner in town—many families would make the trip once or twice a year for checkups or visits to the dentist or eye doctor.

Public water and sewer systems had been installed only in the 1950s, and the town's storm sewers consisted of a number of open ditches that led to the bay. The high tide would come in and help flush them out. When a new high school was built on the south side of town in the 1950s, it was constructed right over one of the ditches. If you were in the school's gymnasium when the tide was right and the fish were running, you could hear the sounds of their flapping tails coming from below the basketball court.

Alaska Avenue was the only paved street in town; the others were gravel. Given the amount of snow in winter, it was impossible to keep the streets completely plowed, so drivers were quite comfortable maneuvering on snow. There were no curbs, and what sidewalks existed were built of wood planks.

Most of the businesses were within a block or two of the Alaska-McKinley intersection. There were two grocery stores, Gilson's and Food Lanes; several banks, including the First Bank of Valdez ("Alaska's oldest bank north of Juneau"); Woodford's, a combination clothing-shoe-drug-furnishings-camera store; a tackle shop for fishermen and a gift and sport store that sold guns and ammunition; and for the women in town a dress shop or two. Valdez had a trailer court and a couple of hotels, including the Switzerland Inn, which had once been a children's home but was now trying to capitalize on the tourist boom. Instead of numbers the rooms were identified by the names of Swiss cities.

Saints and sinners had their choice among a roughly equal number of churches and bars. Of the latter, the Village Morgue, at the waterfront, was perhaps the most unusual; it *had* been a morgue once. During the early gold-mining days, it had been home to a stamp mill for crushing ore. No one was quite sure when it became the town morgue. Because the typically heavy snow cover could make it difficult if not impossible to inter a body in the winter, the morgue had the means to store a corpse until the spring thaw, when a grave could be dug. The deceased was kept in cold storage, hung on a meat hook.

The Pinzon, on McKinley Street, which had a forty-foot-long bar that was said to be the longest in Alaska, was owned by Clinton Egan, known as Truck, a member of a longtime Valdez family that rose to prominence in state Democratic politics. Truck's brother Bill, in fact, was the first governor when Alaska became a state. It was said by more than one politician that the de facto headquarters of the Alaskan Democratic Party wasn't in Juneau or Anchorage but in Valdez, at the Pinzon. A local Democrat, complaining once about the state of the party in town, said that it might attract more members if the meetings didn't take place in a bar.

Among the residents of Valdez, some, like the Egans, were

old-timers. Jack Meals's family had stayed around as well, and his son, Owen, had become a pioneering bush pilot and prominent businessman. He owned the diesel generating plant and the telephone exchange, among other ventures.

But many residents had arrived in town more recently, after the war. Some were getting away from bad times in the Lower 48, and Valdez, literally the end of the road, was about as far as they could get and still be in what passed for civilization. Some had served in the military during the war and having spent time in the town had vowed to return. Valdez had that kind of effect on people—it was a quiet, isolated place, set amid scenery that would take the breath away.

Even with the decline in the shipping business, the waterfront was still a major focus of town life. Where Alaska Avenue reached the water, a long gravel and dirt causeway carried trucks and other vehicles out across the mudflats to Valdez's main dock. Two blocks south, Keystone Avenue led to a shorter causeway and a smaller dock, with a pipeline for unloading fuel oil, which was pumped into storage tanks just onshore. Between the two docks the mudflats had been dredged to build a harbor for Valdez's many small fishing and pleasure boats and the *Gypsy,* and a slip for the state-run ferry to Cordova, which could not be reached by road. The MV *Chilkat,* which could carry fifty-nine people and up to fifteen vehicles, had entered service in the summer of 1963, welcomed on its inaugural visit by a flotilla of most of the small boats, decked out with balloons and streamers for the occasion, a drill team and a ribbon-cutting featuring the governor.

But it was the main dock, built on creosote pilings in about fifty feet of water, where most of the activity took place. The dock was large, with several warehouses built on it. Two of its sides were more than three hundred feet long, so that two freighter ships could tie up and unload at the same time. That seldom happened anymore, but single cargo ships still did come regularly to Valdez. When they arrived, people in town would turn out—some to work, others to watch.

4
CLAM BROTH AND BEER

George Plafker had a problem. He was finishing his second year at Brooklyn College, the public university just off Flatbush Avenue in the borough's Midwood neighborhood, in the spring of 1947, and was all set to transfer to City College of New York, in Upper Manhattan, to complete his degree. He was good at science and math and had decided to major in civil engineering. Anxious to escape New York and his small, cramped apartment, he figured there would always be highways and bridges to design and build around the world. He'd spent the two years at the beautiful Georgian-style campus loading up on all the necessary courses in preparation for the switch to CCNY's engineering school. But now, out of the blue, administration officials examining his transcript discovered that he hadn't taken a required pre-engineering course—Geology 101.

Plafker was frustrated and disappointed. He grudgingly admitted that a geology requirement made sense. After all, to design and build structures you need to know something about the ground they will sit on. But it was the end of the school year, and without that credit he wouldn't be able to start at CCNY in the fall. His plan to escape New York would be temporarily derailed.

There was a solution, his advisers at Brooklyn College told him. The school was offering an introductory geology class in its summer session. There was no one on the regular faculty to teach it, but they'd found a mineralogist with some teaching experience, a rock hound named A. C. Hawkins, to do the job. To Plafker it didn't sound very promising, but on the other hand the course probably wouldn't interfere much with his plans to make some money tending bar at a Coney Island restaurant over the summer. What else could he do, really—he had to take the course if he was going to get his civil engineering degree.

To his surprise, that summer would alter the course of Plafker's life. A. C. Hawkins, the replacement teacher, a genial, pipe-smoking older man who had a passion for rocks and who believed strongly in the importance of seeing them firsthand, would play an important role in shaping Plafker's career—as would generous amounts of free clam broth and cheap beer.

As a Brooklyn College undergraduate, Plafker had the trappings of a New Yorker—the accent and attitude—but he wasn't a genuine native. He'd been born in 1929 outside Philadelphia, in Chester, an industrial river town just west of the city, home to one of the country's largest shipyards. He was the second son of Nathan and Florence Plafker; his brother, Lloyd, had been born two years before.

Nathan Plafker had arrived in the United States about eight years before that, in his early twenties. A native of Galicia, in southern Poland, he had worked for a while in a coal mine in Germany. But with Soviet Russia and Soviet Ukraine battling Polish forces nearby in the Polish-Soviet War, he decided to leave, wary of becoming cannon fodder himself. He stowed away on a steamer bound for Montreal and eventually made his way into the United States.

With two brothers in Delaware, he settled in Chester, just ten miles up the Delaware River from Wilmington, where he worked as a counterman in a deli. For a while he went into the grocery business with one of his brothers. But the two had a falling-out, and Nathan set his sights on New York.

It was on a trip to Brooklyn to see about job prospects that he met his future wife, a Polish immigrant who had come to the

United States several years before. Married in Chester, the couple moved to Brighton Beach after the boys were born. But it soon became apparent that Florence was mentally ill. Diagnosed with schizophrenia, she was institutionalized at King's Park, a sprawling asylum on the North Shore of Long Island, when George was three.

Overwhelmed by having to care for two young boys and struggling financially in the midst of the Great Depression, his father moved the family in with his wife's sister and her husband, Otto and Pauline Glassman, who had a three-bedroom apartment on the other side of Ocean Parkway, in Coney Island. Nathan Plafker rented a bedroom from them, which he shared with the boys.

Since the Glassmans had three young children of their own, and a fourth several years later, conditions in their household were cramped. So in 1937, Nathan sent George and Lloyd to the Hebrew National Orphan Home in Yonkers, just north of the city in Westchester County. George was eight. He managed to maintain his composure until his father left; but afterward he cried, off and on, for several days.

The orphanage, known to everyone who lived there as the H, had been established in 1912 on Manhattan's Lower East Side to provide an Orthodox Jewish upbringing for one hundred needy boys aged six to eighteen. They were housed in two tenements, back to back with a courtyard between, off Second Avenue. In 1920, the home was relocated to a twenty-acre site in Yonkers. Its mammoth four-story main building had dormitories on three floors housing three hundred boys, a dining hall, a synagogue, classrooms and two gymnasiums. The grounds, which backed up against the aqueduct that supplied New York City's water, were bucolic, more rural than urban. In addition to a playground and ball fields, there were fields for growing vegetables, an apple orchard, a chicken coop and a stable with a broken-down racehorse named Tiny Tim. Across the street was a public golf course, where the older boys could work as caddies.

The orphanage functioned with a certain military-like efficiency:

rise-and-shine at 6 a.m., outside for the flag-raising, back inside for prayers at the synagogue, followed by breakfast, marching all the way. Then it was off to Public School 404—the classrooms on the fourth floor, which were under the jurisdiction of the New York City school system. An hour for play or music practice in the afternoon, followed by mandatory Hebrew school, more prayers, then supper, showers, studying and lights-out. There was a certain level of discipline as well, often enforced through what was called detention—being forced to stand or squat motionless for long periods.

Despite the strict regimen and discipline, by most accounts, the H was not an unhappy place. For most boys, there was so much to do—sports, music and clubs, in addition to regular studies and Hebrew—that they didn't spend a lot of time thinking about the circumstances of why they were there. George and Lloyd were more fortunate than most: their father visited them just about every weekend, making the long trek by subway from Brooklyn.

Like everyone else at the H, Plafker studied Hebrew, although it never stuck with him all that well. He chafed at having to attend synagogue. And he didn't much care for some of the orphanage's stricter rules. If his bed wasn't made perfectly, a supervisor would flip the mattress onto the floor. And detention was doled out on the slightest of whims. But Plafker also played a lot of sports, learned how to use a printing press and caddied at the golf course for pocket money. And years before he came to know backcountry Alaska, he explored the wilds of Yonkers, where he and his fellow Boy Scouts set out traplines to catch squirrels and rabbits and went on campouts to the far end of the property.

At age seventeen, Lloyd left to join the navy. George, by then a gangly fifteen-year-old, wanted to get his high school diploma as quickly as possible and go to college for his degree. That, he knew, was his ticket out of New York.

He had been encouraged by his teachers to attend summer school for two years so that he could finish high school early. The summer school, recognizing that George had no money, waived the tuition. And so in 1945, at the age of sixteen, Plafker graduated from Roosevelt High School, moved in with his father in Brooklyn and enrolled at Brooklyn College.

Years later, Plafker would credit the eight years he spent at the H in no small way with shaping his character. Lacking the love and assistance most children receive from their families, boys at the home learned to get along with others, to share, to treat others fairly and to rely on one another for advice and help. Above all, Plafker felt that the home, with its rules, strict discipline and lack of parental authority, instilled in him a kind of inner toughness, a doggedness in dealing with a problem or a task. It was a quality that would serve him well in the years ahead.

Alfred Cary Hawkins, a gentlemanly New Jerseyan, believed that you couldn't learn geology just by reading books or listening to lectures. You had to get out in the field, see rocks in their natural settings, describe and map them and chip pieces off for close-up study and inspection. His conviction that fieldwork was invaluable had informed Hawkins's entire career. Both his master's and doctoral work had involved extensive field studies. (His dissertation, for which he received a PhD from Brown in 1916, was titled simply "The Geology of a Portion of Rhode Island.") After a stint in the army during World War I, Hawkins had taken a desk job analyzing minerals for the DuPont chemical company. But the lure of fieldwork was too strong, so he left DuPont after a few years for a succession of jobs with museums and universities, both teaching and collecting mineral specimens. By the summer of 1947 he was working part time as a consultant and part time as a teacher and hunting for minerals every chance he got, to sell to collectors.

At his first class at Brooklyn College that summer, Hawkins made his students an offer: in addition to the regular course work, they could spend Saturdays with him out in the field, studying rocks up close. Plafker and three other students, intrigued and eager to get out of Brooklyn as much as possible that summer, signed up.

Some of the trips weren't so far afield. Hawkins and his students spent a Saturday in Manhattan, studying the exposed schist of Central Park, bedrock that had been formed 450 million years ago and had been twisted and folded since then. They ventured to the New Jersey side of the George Washington Bridge to look at the later sedimentary rock that overlies the older formations.

But most Saturdays, Plafker and his fellow students took the ferry across the Hudson to the waterside city of Hoboken. Hawkins would drive over from his home in Newark, ten miles away, and off they'd go in his car. One day they explored the Delaware Water Gap, on the Pennsylvania border, where the Delaware River cuts through the Appalachians, exposing gray quartzite. Another trip took them to Franklin Furnace, an old mine in northern New Jersey famous for minerals that glow brilliant colors under ultraviolet light. They went to so many mines and quarries, in fact, that Plafker began to think Hawkins knew every one in the state. He'd probably collected and sold specimens from all of them.

For Plafker, cruising around the wilds of New Jersey was a welcome respite from the drudgery of summer classes in the stifling heat and humidity of New York City. Seeing ancient rocks and formations brought his dry geology textbook to life. But best of all, at the end of each adventure Hawkins would drive Plafker and the other students back to Hoboken, where they'd go to the Clam Broth House, the landmark restaurant on Newark Avenue. Frequented by dockworkers and other blue-collar types, the Clam Broth House was famous for its two large signs in the shape of hands that pointed toward the entrance, and for its free lunch counter. As long as you bought a beer, you could sit at the counter filling up on beans and bread or other simple fare and drinking all the clam broth you could ever want. To the hungry, impoverished students this was a deal too good to pass up. Hawkins would join them, and while they ate and drank he would hold court, telling stories—whether true or not, Plafker couldn't always tell—of much more exotic field trips he'd been on, to mines in Mexico and other far-off places.

To Plafker, desperate to get away from New York, it was eye-opening. He began to seriously consider switching to a degree in geology. That would mean he could continue going to Brooklyn College. The idea of spending the next couple of years at the City University campus up in Hamilton Heights, in northern Manhattan, had always held little appeal, as it would be a nasty commute.

By the end of the summer, Plafker's mind was made up. I'm going to become a geologist, he thought. It seemed like a good life, drinking beer and telling tall tales.

Over the next two years, Plafker immersed himself in his new discipline. The prerequisite courses for a geology degree were the same as those for civil engineering, so now he was taking all geology courses, all the time.

He continued to live at home with his father, but that soon became problematic. Nathan Plafker had gotten a divorce and soon was remarried, to a woman who had two daughters of her own, both younger than George. They moved into the apartment. George couldn't tolerate all the talk and noise that came along with them.

But at that point he'd developed an extracurricular activity that took up a lot of his nonacademic hours: a girlfriend named Ruth Bersin. Ruth had dark curly hair, blue eyes and a seemingly boundless enthusiasm for doing whatever Plafker wanted to do.

At the H, there had never been much opportunity to interact with members of the opposite sex. Ruth was the first woman he had gotten to know to any degree. Their getting together had at first been something of a diversionary tactic, a favor for a friend, who was going out with Ruth's older sister Lydia.

George and Ruth had hit it off, and George had hit it off with Ruth's father as well. Chris Bersin was the live-in superintendent of a six-story apartment building in Manhattan Beach. Knowing that Plafker was not getting along with his father's new wife, he offered him a room in the basement of his building. The place had much in its favor: it was free, it was far from his father's new wife and daughters, and it gave him unfettered access to Ruth.

When Plafker graduated from Brooklyn College in 1949, he was twenty years old. He consulted a listing of civil service jobs and applied for one out West, working for the Army Corps of Engineers in Sacramento, California. The corps, desperate for help, quickly hired him. Although he and Ruth were dating seriously, he went to California on his own in the summer of 1949. He was ready for the kinds of adventures A. C. Hawkins had talked about. The adventure of marriage was a subject for another day.

During World War II, the corps had been kept busy contributing to the war effort, building everything from bridges, roads and military camps in Europe to enormous uranium-processing plants and other facilities that were critical parts of the American effort to make the atomic bomb. With the war over, the corps was turning its attention back to another part of its mandate, building the dams and other structures that would tame some of America's wildest rivers to protect against flooding, allow navigation, generate electricity and provide recreation areas for the country's growing middle class. Plafker was assigned to work on a couple of these projects, including Folsom Dam, a crucial part of what eventually became California's Central Valley Project. His job was to help determine whether the dam site, on the American River some twenty-five miles from Sacramento, was geologically sound. Folsom, of course, was also home to a notorious state prison, and to get to work Plafker had to pass through the prison grounds, always with a police escort. He would later tell friends that he had done time at Folsom Prison.

Later that year he came back East to ask Ruth to marry him. They were married on New Year's Eve in New Mexico on the way back to California.

After another six months with the corps, Plafker began to

realize that despite having an undergraduate degree in the subject he really knew very little about geology. And if he wanted a better job than what amounted to looking at rock cores all day, he needed to further his education. So he quit to enroll at the University of California at Berkeley and get a master's degree.

He lasted only one semester before Ruth became pregnant. Once again he needed a job, to feed his growing family. While he'd enjoyed his time with the corps, it wasn't the kind of fieldwork he wanted to do. But another federal agency offered plenty of the kind of work he was interested in—the US Geological Survey.

The USGS was established in 1879, charged by Congress with "examining the geological structure, mineral resources and products" of public lands. But it had its origins in surveys undertaken more than fifty years before, as the United States was expanding westward following the War of 1812. These early surveys were geared to evaluating the agricultural prospects of new lands. But as the nation eventually grew to be more industrialized, and as the field of geology developed, later surveys focused more on mineral resources.

By the end of the Civil War, American industry had developed to such an extent, and its need for mineral and other resources was so great, that several major surveys of the West were undertaken, including one along the fortieth parallel, the route of the transcontinental railroad. More than any other efforts, these surveys laid the groundwork for the creation of the USGS; its first director was Clarence King, who had led the fortieth-parallel survey.

Since then, the Geological Survey had greatly expanded the scope of its work to include basic functions like topographic mapping (much of the United States was unmapped) and cataloging of water and coal and oil resources. During the world wars it aided the military by concentrating on minerals that were crucial to the war efforts; after World War II it lent its expertise to the vast growth of roads and other parts of the nation's infrastructure as the American economy took off.

Plafker applied to work at the Survey in 1950 and again was

quickly hired. But if he was thinking of a life hacking through brush in the backcountry, that dream was soon dashed. He was assigned to the military geology branch, in Washington. It meant not only moving back East but taking a desk job, itemizing geological features around the world that might be of use to the military as locations for emergency airfields or river crossings. It was as if he were still working for the Army Corps of Engineers. Even worse, he was relying on the fieldwork of other people.

He was desperate to land a field job. He knew the Survey had an Alaska branch, based in San Francisco at the old US Mint building. Its geologists went north in the summer to do fieldwork and came back to California for the rest of the year, writing up their findings and doing other research. Not a bad prospect, he thought. And he knew that the branch was in need of personnel. There was a lot of terrain to be surveyed in Alaska. Plafker applied for a transfer, and after a year sharpening pencils in Washington he headed back West with his family.

5

THE FLOATING WORLD

The first thing one can see about Alaska from a map of the state is this: it's big. From Ketchikan, in the southeastern Panhandle, it stretches more than 1,300 miles north to Point Barrow, on the Arctic Ocean, and nearly 2,200 miles to the westernmost point of the Aleutian Islands, just a stone's throw from Russia. In all, it includes more than 650,000 square miles of territory, more than twice as much as Texas, the second-largest state. When Alaska entered the Union in 1959, the area of the United States grew by more than 20 percent overnight.

If a map gives at least a suggestion of the state's topography, the second thing that stands out is how mountainous Alaska is. There's the Brooks Range to the north, which entered the nation's consciousness when oil companies started drilling on its north slope in the 1960s. Most of the mountains, though, are farther south, and there are a lot of them—hundreds of high rugged peaks, including some that are among the highest and most rugged in the world. Geologists divide Alaska's mountains into ranges, and while there might be arguments over precisely where one range starts and another begins, or whether one is more accurately a subrange of another, here's one way to sort them:

Coast Mountains
St. Elias Mountains

Wrangell Mountains
Alaska Range
Kuskokwim Mountains
Chugach Mountains
Talkeetna Mountains
Kenai Mountains
Aleutian Range

The big picture, however, is more important, and it's a distinctive one. These mountains scribe a long arc that follows the coastline, starting with the Coast Mountains in the southeast, trending northwest through the St. Elias chain and the Wrangells, making a tight turn in the Chugach and the Alaska Range and trending southwest in the Kuskokwims, the Kenais and finally the Aleutian Range, which ends at the first of the islands at the western tip of the Alaska Peninsula.

The map doesn't show it, but many of Alaska's mountains are volcanoes, active or extinct—peaks like Edgecumbe in the southeast, Churchill and Drum in the Wrangells and dozens in the Aleutian Range and farther west among the Aleutian Islands themselves. The Geological Survey, which has offices in Alaska to monitor and study the volcanoes, says that more than 130 have erupted in the past few million years. About 50 of these have been active in the two and a half centuries since Alaska was first discovered by outsiders. Captain Cook himself observed Mount Redoubt, in the Aleutian Range, spewing white smoke back in 1778. It has had full eruptions four times since 1900, including one in 2009. In 1912, Alaska was home to the most powerful eruption of the twentieth century, at a previously unknown volcanic vent called Novarupta on the Alaska Peninsula. Ash flowing from the eruption covered a nearby valley up to seven hundred feet deep, and for the next fifteen years steam vented from countless openings in the ash layer. Robert Griggs, leader of a National Geographic Society expedition to explore the devastation, named the eerie landscape the Valley of Ten Thousand Smokes.

But Alaska is just one part of what is, in effect, a rim of mountains and volcanoes around the Pacific Ocean, what is known today as the Ring of Fire. They run south through the Pacific Northwest of the United States, with Washington's Mount Rainier being the best-known volcano, and on to California, Mexico (including Popocatépetl, outside Mexico City) and Central America. From there they continue into South America, where the Pacific coast is characterized by long chains of peaks from Ecuador all the way to Chile. Across the Southern Ocean, the mountains continue in Antarctica and then north along the western edge of the Pacific, from New Zealand to Indonesia—home to the famous Krakatoa eruption of 1883—and the Philippines (Mount Pinatubo, 1991) and on to Japan and the Siberian coast of Russia.

There are mountains and volcanoes elsewhere in the world, of course, but these stand out because they are so neatly arrayed around the Pacific. Clearly, for all these mountains to have formed, something—some process involving the outer layers of the earth—must have happened at the margins where the oceans and continents meet. And the volcanoes, just as clearly, suggest that that something is still going on.

It took a long time for science to figure out what that something was. As it turned out, events in Alaska would play a crucial role.

In the late 1940s, when he was a distinguished scientist looking back on four decades of accomplishments, the German geologist Hans Cloos would recall the day in 1915 when he first met Alfred Wegener. Cloos, who was twenty-nine at the time, had taken a teaching position at the University of Marburg, in central Germany, after a few years in Asia working for an American oil company. Soon after his arrival, as he later wrote, "A man came to me, whose fine features and penetrating, gray-blue eyes I was unable to forget."

Wegener, who was five years older than Cloos, was a lot of things—physicist, astronomer, meteorologist, teacher, explorer—

but he was not a geologist. Born in 1880 and raised in Berlin, the son of a theologian, Wegener had studied at universities in Germany and Austria and earned a doctorate in astronomy in 1905. A decade later, Wegener, who by then was concentrating on climatology and meteorology, had already achieved much. He had become an expert on the use of kites and balloons to take atmospheric and celestial readings at high altitudes. He'd taken part in two expeditions to the remote Greenland ice sheet, conducting the first extensive meteorological studies in the Arctic. He had been appointed as a tutor at Marburg and had published a textbook, *The Thermodynamics of the Atmosphere,* that was already a must-read among meteorology students. But when he came to see Cloos, it was something that he'd started working on several years before—something far afield from any of his formal studies—that he wanted to talk about.

"He had developed an extraordinary theory in regard to the structure of the earth," Cloos wrote. "He asked me whether I, as a geologist, was prepared to help him, a physicist, by contributing pertinent geological facts and concepts."

The "extraordinary theory" concerned an age-old question: how the mountains and, by extension, the continents had come to be, or, as Wegener looked at it, how the continents had come to be where they are. For he had developed a seemingly wild idea: that the earth's landmasses were not fixed but instead moved through the oceanic crust. His concept came to be known as the theory of continental drift.

The first spark of the idea had occurred to Wegener in early 1911, when a colleague at Marburg had shown him a new atlas he had received for Christmas. Looking at a map of the world spread across two facing pages, Wegener noticed something about the continents of Africa and South America: their coastlines, with all their twists and turns, were parallel to each other, although they were separated by thousands of miles of ocean. If one could put the west coast of Africa and the east coast of South America together, they'd fit practically like two pieces of a puzzle.

This was hardly a unique revelation. Scientists and others had noticed the similarities between the two coastlines since at least the sixteenth century (and it has been apparent to generations of schoolchildren in more recent times). Wegener went a little further. The map he was looking at was a new one, with depth profiles of the ocean floor off both continents, showing the contours of the continental shelf off each coast. These, too, matched quite well. The congruence of the coastlines, then, wasn't some artifact of erosion or other effect of the oceans.

To Wegener, what he saw could have only one meaning: the continents were somehow able to move. He could see enough evidence from other parts of the world—the way Europe and North America appeared to fit together, for example, particularly if Greenland was included as a kind of wedge—to think that similar movements had occurred everywhere. He realized, too, that this might explain how mountains arose. Perhaps, as the continents had moved, their leading edges had become distorted and pushed up.

Wegener had quickly put all these thoughts aside, however, for he had too much else going on, scientifically and otherwise. He had to prepare for one of his Greenland expeditions, which would leave whenever he and his fellow explorers could raise the money for it. And in the middle of 1911 he had to drop everything and spend two months in reserve military training, as the clouds of war were beginning to gather over Europe.

Wegener was a generalist with a nearly insatiable curiosity, a voracious reader often in fields outside his own. That fall, back from military training, he'd come across an academic paper that had rekindled his interest in his idea. The subject of the paper was geology, and Wegener understood it well enough to immediately draw a connection to his thoughts about continental movement. The paper detailed the strong evidence, from identical limestone formations found in Africa and South America, that the two continents had once been joined.

With the Greenland trip looming, Wegener put his other work on hold for a few months and furiously plunged into researching his idea. He found other evidence of rock formations that seemed to have formed when the continents were joined: research linking coal deposits in Britain and America, for instance. There were also fossil studies, some of which had been around for decades, which showed that the same species of plants and animals existed on continents that were now thousands of miles apart. One was *Glossopteris,* actually a group of similar fern species, now extinct. *Glossopteris* fossils had been found almost everywhere, it seemed, including Africa, Australia, South America and the Indian subcontinent. Other fossil evidence suggested that continents had gone through sharp swings in climate—that what had once been the tropics, for example, was now at a more temperate latitude. This too could be explained if the continents were moving about the planet.

In just a few months of hectic work, Wegener had written a long treatise, "The Geophysical Basis of the Evolution of the Large-Scale Features of the Earth's Crust (Continents and Oceans)," outlining his idea. He'd given a lecture at the Geological Association of Frankfurt and another back at Marburg.

But the reaction had not been particularly enthusiastic (Wegener used the word *indignant* to describe the reception at Frankfurt). And at any rate, Wegener's regular academic work had intervened, as had a second expedition to Greenland. He left for Greenland (by way of Iceland) in 1912 and returned the following year. Then, in the summer of 1914, as war was about to be declared, his reserve unit was mobilized. He joined troops that were invading Belgium. The first time he saw action, in late August, he was shot in the arm. After he recovered and was sent to the front in France, he took another bullet, this time in the neck. His career as a soldier over, he was sent home to Marburg.

In 1915, Wegener took up his theory again, working on a more complete description of it in a book that would end up being titled, more succinctly, *On the Origin of Continents and Oceans.* Cloos, he figured, could help him get the geology right.

Over the next two months, the two collaborated. Cloos wrote later that he never fully subscribed to Wegener's thesis, which "loosened the continents from the terrestrial core, and changed them into icebergs" of rock floating on oceanic crust. But the men became friends, and their constant debate and discussion helped Wegener sharpen his ideas.

Wegener saw signs of continental drift all over the globe, including the way the Arabian Peninsula appeared to be moving to the northeast. Cloos had traveled through the Red Sea and the Suez Canal on his way to and from Asia, and in one of their discussions Wegener had appealed to his knowledge of the region's geography. Wegener's words, as related by Cloos, show how passionate he was about the subject.

"'Just look at Arabia!' Wegener cried heatedly, and let his pencil fly over the map. 'Is that not a clear example? Does the peninsula not turn on Sinai to the northeast like a door on a hinge, pushing the Persian mountain chains in front of it, attaching them on the two hooks of Syria and Oman like drapes!'"

In the nineteenth century, one of the prevailing ideas of earth science was that the planet was cooling. In this view, the earth had formed as a hot ball and had been losing heat ever since. Like many hot things—a pair of blue jeans just out of the dryer, say—as it cooled, it shrank. Eduard Suess, an Austrian geologist, was among those who developed this theory of contraction, as it was called, and had come up with an analogy for it that made it easy to understand. The cooling earth, he said, was like a drying apple. As the flesh of the apple shrank, the skin became wrinkled. As the earth's interior cooled and contracted, its crust folded and wrinkled, forming mountains.

Suess thought that something else happened as the earth cooled: parts of the crust sank, forming the ocean basins. The parts that didn't sink were the continents. But he went further, arguing that as cooling and shrinking continued, eventually the landmasses, in turn, would sink below the level of the ocean basins. The continents would become oceans, and the oceans, continents. That would help explain why fossils of marine organisms were often found on what was now dry land, for instance.

Contraction theory had been widely accepted, but by the turn of the twentieth century there began to be some gaping holes in it. Radioactivity had been discovered in the mid-1890s, and scientists soon realized that the decay of radioactive elements produces heat. Radioactive elements like uranium are abundant in the earth (it was uranium, in fact, that had led Henri Becquerel to discover radioactivity in the first place). So how could the planet be cooling if there was a constant source of new heat deep within it? And if the planet wasn't cooling, how could it be shrinking?

There were other problems with contraction, most notably what was being learned about the earth's crust through gravity measurements like those undertaken in India in the mid-1800s. Crews working for Colonel George Everest, the surveyor general of India (and the man for whom the highest mountain in the world was later named), showed that the continents were less dense, or lighter for a given volume, than the material below them. It's as if a continent were an ice cube: just as an ice cube floats in a glass of water (because ice is less dense than water) and if pushed to the bottom will pop right back up, it seemed that the continents were "floating" on denser material and that it was impossible for a continent to sink and become an ocean basin.

The gravity findings supported Wegener's ideas. He had argued that the continents didn't move vertically anyway, but rather horizontally. (In his original treatise and book, he referred to "horizontal displacement" rather than drift, though they meant the same thing.) If the continents "floated" on oceanic crust, it was easier to envision them moving around. The movement was no doubt very slow, taking place over millions of years. It was hardly perceptible, yet Wegener was convinced by the evidence that it had happened.

To have drifted apart, the continents had to at one time have been together. Wegener borrowed a concept that Suess had developed, that the continents at one point formed a giant landmass, a supercontinent of sorts. Suess had called it Gondwanaland. Wegener called it Pangaea.

Wegener's book, a slim ninety-four-page volume, was published in 1915; he revised and expanded it several times into the 1920s.

The gist of his theory was this: about 225 million years ago, all the land on earth had been part of Pangaea. Then landmasses began to drift apart. As they moved, the leading edges crumpled and formed mountains; that explained why many ranges are found on coasts. Some landmasses would collide and become one, which explained why some mountains were in the interiors of continents. Over time, most of the landmasses had ended up far apart. They were the continents we know of today, and the spaces between them were the oceans.

Wegener cited the various geological and fossil findings that supported his idea. He also cited the gravitational work that showed that the continents could not sink, and he buttressed this with the fact that the continents were largely made of granite, a less dense, and therefore lighter, rock than the basalts that made up the ocean floors. So it made sense that the lighter continents could move across the heavier ocean basins.

All of this would occur very slowly, Wegener argued. Continents moved on the order of inches a year. But with enough time, even tortoise-like movements could result in big changes. Wegener theorized that the drifting had occurred over tens of millions of years and was still occurring.

Like his 1912 treatise, Wegener's more complete theory of 1915 met with much skepticism. He was widely ridiculed and mocked, in part because he, as a nongeologist, was attempting to tell geologists how the world worked. But it was in 1922, when his book was finally translated into several languages, including English, that the loudest criticisms began.

Some of the critiques of continental drift stemmed from Wegener's proposed explanations of how the continents moved. He argued that the forces created by the rotation of the earth could make the continents move slowly across the oceanic crust. Alternatively, the gravitational pull of the sun and moon might provide the motive power. Both of those mechanisms were easily dismissed by scientists with more knowledge of physics than Wegener.

But much of the rejection of Wegener's theory focused more on his methods, and on the fact that he was a geological outsider. This sentiment was particularly felt in the United States, where, in 1926, the American Association of Petroleum Geologists organized a symposium in New York City to discuss continental drift. Held at the stately Pennsylvania Hotel, across the street from Pennsylvania Station on Seventh Avenue, the event allowed some of America's leading earth scientists a chance to sound off about the theory. It was more a kangaroo court than a symposium. Edward Berry, a professor at Johns Hopkins University in Baltimore, attacked the method by which Wegener had arrived at his theory, which, he said, "is not scientific, but takes the familiar course of an initial idea, a selective search through the literature for corroborative evidence, ignoring most of the facts that are opposed to the idea, and ending in a state of auto-intoxication in which the subjective idea comes to be considered as an objective fact." The paleontologist Charles Schuchert, who had headed Yale's natural history museum and who knew as much about the distribution of fossils around the world as anyone, accused Wegener of being far out of his depth. "Facts are facts," Schuchert wrote, "and it is from facts that we make our generalizations, from the little to the great, and it is wrong for a stranger to the facts he handles to generalize from them to other generalizations."

The world of geology was being divided into two camps in relation to what had come to be called "tectonics"—the processes that affect the earth's crust. One, the mobilists, accepted the basic premise of Wegener's theory, even with its flaws. The other, the stabilists (sometimes also known as fixists), stuck to the older contraction theory or some updated variant of it (Schuchert himself at one point proposed the idea of "land bridges" that connected the continents and came and went). By the mid-1920s, the stabilists were prevailing, and Wegener was suffering for it. He had been hoping for a full professorship at Marburg or another university in Germany but was continually turned down for advancement. It was not until 1924 that he received a faculty appointment—in meteorology, in Austria, at the University of Graz.

Wegener published a new edition of *On the Origin of Continents and Oceans* in 1929, and in the next year he left on his third expedition to Greenland. Where the first two expeditions had been organized and led by Danes, this time Wegener was the leader. The main goal was to set up permanent weather stations on the ice sheet. But there were problems with logistics, and after journeying to the middle of the ice in late fall to resupply one of the stations, Wegener and another member of the expedition set out to return to their base camp. They never made it back. Wegener's body was found the next year; he had apparently died of heart failure along the way. He had just turned fifty years old.

Wegener died knowing that his theory was largely discounted within the scientific community. In the world of earth science, stabilists were to rule the day for the next three decades. But beginning in the 1940s, new evidence was uncovered from beneath the oceans that supported the idea that the continents moved—though not in the way Wegener had thought.

War may be good for very little, but occasionally in the world of science, at least, it can be helpful by spurring discoveries. That was the case during World War II, when, as a side benefit to the American military campaign to beat back the Japanese in the Pacific, a scientist named Harry Hess laid the groundwork for a fuller understanding of the subject that Wegener had explored years before.

At the time, late in the war, Hess was the captain of the USS *Cape Johnson,* a navy attack transport that delivered troops for landings at Iwo Jima, Leyte and other locales in the Pacific theater. The *Cape Johnson* was equipped with a powerful fathometer, a sonar device that pings sound waves off the ocean floor, using the reflections to measure depth and create a profile of the bottom. Depth finders like this were useful to the military, especially for making charts of shallow waters that showed reefs or other hazards. Hess dutifully used his fathometer in that way, but he did something else—he left it on all the time, even when he was far from land.

Hess was no ordinary naval captain. Raised in Asbury Park, New Jersey, and a 1927 graduate of Yale, he'd received a doctorate in geology from Princeton in 1932 and had taught there and at Rutgers before the war. He had also participated in a field study in the Caribbean, making gravity measurements from a navy submarine. For the submarine work he'd had to become a navy reservist, so when the Japanese attacked Pearl Harbor in 1941 Hess reported for active duty the next day. He spent a couple of years analyzing the patterns of German U-boat movements in the Atlantic, helping the Allies to all but eliminate the threat from the submarines, before shifting to the Pacific.

Leaving the fathometer on day and night as the *Cape Johnson* steamed along—even diverging slightly from its assigned course at times, in the name of science—Hess obtained contour maps of a sizable portion of the North Pacific. The maps revealed that the seafloor was littered with features, including canyons and trenches and, most strangely, flat-topped mountains that rose from the seafloor yet ended hundreds or thousands of feet below the ocean surface. Hess counted more than 150 of them and named them guyots, after Princeton's first geology professor. To Hess, these were clearly old volcanoes that had had their tops eroded by the ocean. Together with what he'd learned from the work in the Caribbean, which showed that extremely deep parts of the ocean had lower than normal gravity, the nature of the oceanic crust was beginning to be revealed. Rather than a vast boring expanse of rock, the seafloor, like the continents, showed signs of geological activity. It would take more than a decade, and more detailed mapping and other studies by a number of researchers, for Hess to propose what that activity was.

Some years before Hess's fathometer work, another scientist had speculated—quite accurately, as it turned out—as to what might be going on. Arthur Holmes was a British geologist, born in 1890, who as a teenager had been inspired to study the subject by reading the works of Eduard Suess, the contraction theorist. Early in his career he had become interested in radioactivity, and he was a pioneer in the development of radioactive dating, using the rate of decay of isotopes in rocks to determine their age. At age twenty-three he had published a seminal work, *The Age of the Earth,* in which he estimated that the planet was 1.6 billion years old.

On the question of the continents, Holmes at first had adopted Suess's ideas about a cooling, contracting earth. But as the years went by and he learned more about radioactivity and heating, he eventually rejected contraction in favor of continental drift. Holmes thought a lot about Wegener's theory, focusing on what most people considered its weakest link, the lack of a suitable mechanism by which the continents would plow across the oceanic crust.

By the mid-1920s, scientists had some understanding of the earth's interior, although it was still limited. There had been a series of discoveries based on the study of how waves produced by earthquakes were affected by traveling through the earth, starting with Richard Oldham, a British geologist who, in 1906, found evidence that the earth had a core. Three years later a Croatian, Andrija Mohorovičić, discovered a boundary between the planet's top layer, the crust, and a region below it, called the mantle. This boundary varied in depth, from less than ten miles below the ocean floor to as much as sixty miles below the continents. Finally, in 1915, the German geologist Beno Gutenberg (who went on to develop the earthquake magnitude scale with Charles Richter) found what appeared to be a boundary between the core and the mantle, at a depth of about two thousand miles.

One big question was whether this large mantle region was liquid, as recent work had suggested that at least part of the core was, or solid, like the crust. Holmes, through his knowledge of radioactive decay, knew that there was a lot of heat inside the earth. In *The Age of the Earth* he'd suggested that temperatures could reach 1,300 degrees Fahrenheit or higher not too far down. Perhaps, he thought, the mantle was basically an extremely thick liquid. Mostly it behaved like a solid, but under certain conditions it could slowly flow like a liquid. There was plenty of heat down there, and it had to go somewhere. Perhaps there could be convection within the mantle.

Convection is all around us. It's what makes water heat up in a pot on the stove, for example. As the water warms from the heat source below it, it expands, becomes less dense and rises. As it does so, it displaces colder water at the top, which descends because it is denser. Now this colder water is closer to the heat source, so it warms and rises, and the warmer water at the top, now farther from the heat source, cools somewhat and falls. A current is created as the water goes round and round, rising and falling as it heats and cools; the zone where all this is occurring is known as a convection cell. Since heat is continuously being added at the burner, eventually the entire pot of water is hot.

Convection can occur in any liquid or gas. Hot air rises and colder air falls in the atmosphere, creating winds and, ultimately, weather. Cold water falls and warmer water rises in the ocean, and together with winds creates currents. And, Holmes suggested in a paper published in 1929 by the Geological Society of Glasgow, convection could set up currents in the mantle. Hot viscous rock would rise, spread out along the top of the mantle and then fall back down as it cooled. Most important for drift theory, Holmes argued that convection currents could be the mechanism driving it. The pressure of the rising mantle rock could cause a continent to break apart, with the pieces spreading to either side.

The idea was intriguing, but it was also largely ignored. Holmes himself apparently did not spend a lot of time defending it. Much later, in writing about it, he described his idea as speculative and added that "purely speculative ideas of this kind, specifically invented to match the requirements, can have no scientific value until they acquire support from independent evidence."

By the late 1940s that independent evidence was starting to come in, as more scientists took up the study of the ocean floor. Bruce Heezen, of what was then known as the Lamont Geological Observatory at Columbia University in New York, led an effort to map the North Atlantic aboard a research ship, the *Vema,* which was equipped with a highly accurate deep-ocean fathometer. Back in New York, a Lamont geologist and cartographer, Marie Tharp, drew maps from Heezen's data. (The observatory's director, Maurice Ewing, had decreed that women were not allowed on the *Vema.*)

Tharp produced her first map in 1952 and continued making maps for years as more data came in from the Atlantic and elsewhere. (She finally got to go on a research cruise, one run by Duke University, in 1965.) One thing the maps showed was that at the edges of the oceans there were often long trenches that were far deeper than the surrounding seafloor. But the biggest surprise was what the maps revealed about the middle of the oceans.

It had been known since ships began laying telegraph cables between North America and Europe in the second half of the nineteenth century that there was a ridge of undersea mountains in the mid-Atlantic. Now Tharp's detailed maps revealed that the ridge was actually two parallel ridges, with a valley between. What's more, this double ridge extended from north to south for thousands of miles, like a giant surgical scar. And when Tharp mapped other oceans, it became apparent that the mid-Atlantic ridge was just a part of a system of connected midocean ridges that looped around the world.

All this time, Harry Hess, now chairman of the Princeton geology department, had been continuing his own research and following closely the work of Heezen, Tharp and others. He looked at the evidence—the ridges, with their valleys, in the middle of the oceans, the trenches at their margins, and his own findings of guyots in the Pacific—and in a great example of scientific synthesis came up with a hypothesis to explain them. What if, he proposed in 1959, the midocean ridges were places where hot rock, or magma, was welling up from the mantle by convection, oozing out to become the seafloor and spreading apart. In his view, the valleys found between the twin ridges were evidence of this spreading, or rifting. At the ocean margins, the trenches were areas where the spreading seafloor, being denser than the continents, sank down again into the mantle, a process that came to be called subduction. The mantle thus contained convection cells that were like conveyor belts, bringing magma to the surface at the ridges (Hess referred to these as "ascending limbs") and carrying it along to the edges of the oceans, where it went back down to the mantle ("descending limbs") and eventually back to the ridges. In a way this was similar to what Holmes had proposed years before, but Holmes had suggested that the magma welled up under the continents, not at the ridges.

Hess wrote that the process was extremely slow, with the seafloor moving about half an inch a year. But his idea, eventually known as seafloor spreading, explained a few other things. As he wrote later, it provided "a more acceptable mechanism" for continental drift, "whereby continents ride passively on convecting mantle instead of having to plow through oceanic crust." But the sinking of the oceanic crust back into the mantle at the edges of the oceans could also help account for the mountains that are found all along the edges of the continents. The continents, he wrote, "are strongly deformed when they impinge upon the downward moving limbs of convecting mantle."

It was a strong hypothesis, but, as Hess himself acknowledged in a paper describing it in 1962, it was little more than an "essay in geopoetry." More evidence was needed to confirm that the seafloor was really spreading, and in the way Hess described. That evidence would soon be forthcoming.

6

SPIKING OUT

The grizzly came rushing out of the alders on the other side of the river, headed straight for him.

George Plafker had been working for the Alaska branch of the Geological Survey for a couple of years, and during that time had spent weeks in the bush. You can't spend as many days in the Alaskan wilderness as he had and not be familiar with bears. Plafker had come across both grizzlies and black bears before, on the trail and near his camps. His standard procedure in such cases was to speak up and back off at the same time. He'd shout or speak loudly, perhaps saying something like "Okay, I'm backing away as fast as I can." Usually the sound of his voice was enough to set the bear scampering in the opposite direction. Occasionally he'd have to grab his gun and shoot in front of the bear's feet—send some gravel up its butt, as he'd put it—to get it to hightail out of there.

But this time, he didn't have a chance to react. Plafker and an assistant were out working in the Samovar Hills—an isolated chain of short peaks in the southeast, on the edge of the Malaspina Glacier, one of the largest in Alaska. Plafker was wearing his usual khakis, and his head sported a bright red cap. Like everyone else in the Survey who did fieldwork, he made it a point to always wear something red. That way, if he was working by himself and keeled over or was otherwise incapacitated, his colleagues had half a chance of finding him.

It was late in the evening, and dusk was coming on slowly, as it does in the Far North, where the sun sinks obliquely below the horizon. Plafker had been anxious to get back to camp before nightfall fully set in. He was hiking along what was obviously a bear trail—the only trail of any kind in this remote area—at a steady clip. His assistant, who was carrying Plafker's .30-06 sawed-off rifle, was dragging a bit and was some distance behind. Plafker decided to wait up for him and take back the weapon. It was a good thing he did.

Having resumed hiking and once again comfortably ahead, Plafker came upon a small stream, about fifty feet from bank to bank. The water coursed by in shallow rivulets, leaving plenty of gravel bars that would make crossing easy. The trail appeared to continue on the other side.

Plafker started across, onto one of the bars, when he saw movement on the opposite bank. He wasn't sure what it was, but he could hazard a guess. He stopped, then backed up a bit. That's when the bear charged.

Plafker didn't have time to raise the rifle; he fired once from the hip. He didn't appear to hit anything, but the noise and flame caught the grizzly's attention. Startled, it stopped and turned sideways. By then Plafker had reloaded, put the rifle to his shoulder and fired another round. The bear dropped dead.

A few moments later his assistant emerged from the trail, looking around. "What are you shooting at?" he asked.

Afterward, Plafker considered that perhaps the animal hadn't meant to be aggressive. Maybe it was just cornered, he thought—the terrain on the other side of the stream was steep, so perhaps the bear had had nowhere to go but across the water toward him. Perhaps if he hadn't shot at it the grizzly would have raced right by him on the trail. But then again, if the bear had raced right by him and while doing so had happened to take a swipe at him with its claws, Plafker's head would probably have been ripped off.

This was an element of the Alaskan bush that Plafker hadn't thought much about when he had sought a transfer to escape a desk job with the Geological Survey in Washington. But it added a certain amount of exhilaration and adrenaline to the work.

Not that he needed much more exhilaration. It had taken only a brief time working in Alaska for him to realize that he'd made the right decision to come here.

He and Ruth and their infant daughter, Linda, had headed back to the West from suburban Washington in the spring of 1952. They'd settled first in an apartment and later, with $500 from her parents, had bought a house south of San Francisco. Shortly after he arrived at the Survey, it was fieldwork season in Alaska. So he left Ruth and the baby and headed north.

At the time, the primary job of the Alaska branch was to learn enough about the geology of the state that its resources could be exploited. Some of those resources were metals like gold, copper, nickel and platinum. And increasingly, some were coal and oil.

The USGS had plenty of topographic maps of Alaska, with their familiar whorls showing elevation contours and details like roads, streams and mountain summits. Since the 1930s, these had been made using aerial photography. But to determine the kinds of rocks that make up a given piece of land, how those rocks came to be there and what minerals or other valuable resources they might contain, most of the work has to be done on the ground. Those doing it need to have expertise. They need to be geologists.

Geologic mapping is old-school science, about the most basic work that a geologist can do. You hike across terrain—preferably in a systematic way—looking for exposed rock formations. When you find one, you note its location and what kind of rock it contains—an old igneous rock like granite, say, or younger shales or sandstones or other sedimentary rocks. If you find something that's particularly intriguing—embedded minerals that you aren't sure about, or some fossils—you take out your rock hammer and chip off some samples to take back to the office.

You examine the rock formation to see what has happened to it over time—signs that it has been pushed up, folded over or broken apart. If the formation is tilted, you measure its angle of incline (called "dip") and the dip's compass direction (called "strike"). You note its precise location in your field notebook and on your topo map—as precise as can be determined—and sketch its extent, if possible. Then you keep hiking in the same direction until you come upon more rock, and do the same thing. Later, back in the office, you map all of this, using your expertise to fill in the blank areas between the formations you saw.

Plafker loved the work, and like most geologists he got better at it the more he did it. He learned that creeks were often the best place to explore because over time the water had cut through the landscape, exposing layers of rock on either side. Alaska had no shortage of creeks, and by hiking up one and noting the formations, then hiking across to the next creek and back down it, he could fairly easily construct a geological map of an area.

At first Plafker had worked as an assistant, learning from Survey geologists with more experience and expertise. There were some characters among them, men like Darwin Rossman, whom everyone called Hardrock because of his penchant for studying granites and other rocks that had been a product of heat and pressure within the earth, rather than softer sandstones and the like that resulted from the buildup of sediments. Another was Don J. Miller, who had been mapping in the Chugach Mountains and elsewhere in southern Alaska for years. He became a mentor of sorts to Plafker, initiating him in the ways of the field.

Miller, ten years older, had been obsessively interested in rocks since his childhood growing up on a farm outside Akron, Ohio. He'd joined the Alaska branch in 1942 straight out of the University of Illinois, where he'd earned a master's degree. After a couple of years of investigating copper and nickel deposits, he'd been assigned to the group looking for signs of oil.

A geologist who studies rocks for their potential to contain oil reserves is of necessity looking at sedimentary formations.

These rocks begin as deposits of sediment washed down over time from the weathering or erosion of other rocks, or as layers of dead organisms or other organic matter that settles out of a body of water. The sediments eventually become compacted and hard, and trapped organic matter decays to oil or gas. Petroleum geologists are also interested in how sedimentary rocks are folded or otherwise deformed. Pressure could cause them to wrinkle, like a tablecloth pushed from one end. The tops of these wrinkles—called "anticlines"—can trap oil and gas.

Some anticlines are easy to see. It's been said that when petroleum geologists were first exploring for reserves in Saudi Arabia, they could spot promising anticlines in the desert landscape from far away. But in most cases, anticlines or other oil-trapping features are located deep underground, and the surface topography doesn't necessarily provide clues. So scientists like Miller had gotten good at studying the layers of sedimentary rocks—analyzing the stratigraphy—and how their folding might suggest the presence of oil.

Plafker did a lot of different things in those early years in Alaska. For one thing, he surveyed four potential sites in the south-central part of the state for hydroelectric dams. But he also found himself exploring the sedimentary rocks of southern Alaska with Miller.

The work would begin earlier in the year in the office in Menlo Park, studying topo maps and any mapping that had already been done in a given area. The goal was to identify places to study and map those that were potentially interesting and logistically possible to reach. Once in Alaska for the field season they would use Cordova or another town as a base and make forays into the field for up to a week at a time—"spiking out," they called it.

Plafker came to greatly appreciate the bush pilots who took him into the backcountry. They were an unusual breed. Some people might say they were crazy and, at times, reckless, but Plafker saw them as being good at problem solving and improvisation. It just so happened that they did much of their problem solving and improvisation when taking off or landing in places where most people would not think it was possible to put a plane.

Plafker had been on many flights where the pilot had set his plane down in the middle of one of Alaska's wide rivers, on a gravel bar. These bars often had logs or other debris on them, but the pilot would pick out a place that was clear enough for landing, and the plane usually had large low-pressure tundra tires that could handle rough terrain. Leaving, though, was another matter, because the plane required a longer run to take off. So after unloading, Plafker and the others would spread out over the bar and clear enough debris to make a runway that was long enough.

Problem solving and improvisation sometimes led to misadventures. Plafker had been at a cabin near a lagoon along the coast once when a plane equipped with both skis and wheels came in to pick him up. The pilot thought he'd be able to land on a mudflat at the edge of the lagoon, but it was too small. So he came in on the water with his skis down, keeping his speed up enough so that he skimmed across the surface like a water skier until he reached the mudflat. It worked for the landing, but taking off was trickier, especially with a full load of people and supplies. The pilot had to reverse the process—begin the takeoff on wheels on the mudflat and then, when he reached the water, jack the skis into position at just the right moment, hoping that he had enough speed to stay afloat and eventually take off. It didn't work the first time—the plane didn't sink, but got awfully wet before they were able to drag it back to shore—and they had to remove equipment to lighten the load until the technique worked. Plafker found that in situations like these the tension level would increase with each failed takeoff.

Most flights were far less stressful. The pilot would land the plane, the geologists would unload their equipment and supplies, arrangements would be made to be picked up at the same spot in a week or so and the pilot would take off for Cordova, or wherever he was based. But the plane's departure was often the moment during the whole adventure when Plafker felt the greatest unease. That pilot, he would realize, was the only person in the world who knew exactly where they were. If he crashed before he got back home, that knowledge died with him. Sure, they could use the radio to request a rescue, but would they be able to describe where they were? Plafker would put those concerns out of his mind, though, and get to work setting up the spike camp.

A typical two-person camp consisted of two straight-sided military-style tents, one for sleeping and one for cooking. (The tents were often military-surplus items.) To save weight, they never brought tent poles. Instead, they'd cut willow branches and lash them together to make tent frames. It could be tedious work, especially when it was rainy and cold. Then they would set up the two-way radio, often an even more tedious process. The radio had a long antenna, a wire that had to be strung from tree limbs, if possible, for best reception.

The men slept in sleeping bags on air mattresses, ate meals heated on stoves fueled by white gas and had gas lanterns to see by at night. Each morning they would venture out carrying a belt with the usual geology tools, including a rock hammer, compass, level and notebook. But they'd also carry a heavy backpack stuffed with other equipment, food and safety supplies like emergency blankets and signal flares. And they'd carry a weapon for protection against bears. Plafker sometimes hauled fifty pounds in his pack, and throughout the day the load tended to get heavier as he chipped off rock samples to take back.

They'd head off mapping each morning, working together or separately, depending on the circumstances. The idea was to make forays as far as they could go and safely return by the end of the day. When the area had been fully surveyed, it would be time to break camp, call in the airplane and wait for the familiar drone of its engine coming back to pick them up.

Some survey geologists didn't really care what they ate or how comfortable their camp was. Others paid more attention to the details, especially the food. Plafker was one of the latter. While hardly a gourmand, he made the most of mealtime. His favorite backpacking food was the military C ration, which the Survey bought in bulk on the cheap. It always seemed that the rations were five or ten years old—perhaps that was one reason they were so inexpensive, Plafker thought—but as the food was military grade, age didn't seem to affect it. He always brought a large bottle of Tabasco with him into the field to improve the taste.

The thing about C rations was that they were totally self-contained. That appealed to Plafker's sense of order. The army thought of everything. The rations included some kind of canned main dish—pork and beans, maybe, or spaghetti and meatballs—hard crackers, a little container with a buttery spread, even a dessert of some kind, a thick cookie or fudge-like substance. They also contained powdered coffee, sugar and powdered creamer, matches and a wad of toilet paper. Some of them even had four-packs of cigarettes. Those were especially useful for trading with the bush pilots, almost all of whom smoked. Plafker found he could easily swap his cigarettes for an extra dessert.

The main dish could be eaten cold, if necessary, but Plafker was soon initiated into the ways of heating it up. Later in his career, when the Survey started using helicopters to get into the backcountry, heating the food was relatively simple as long as the chopper was around. You'd take an oil filter wrench—basically a band of steel attached to a long handle—wrap it around the can of pork and beans or whatever and hold it in the helicopter engine's exhaust for a minute. Then you'd turn it around and hold the other end in the exhaust for another minute. That was it; your food was piping hot. Timing was everything, though. Keep it in the hot exhaust for too long and the can was liable to explode, sending hot beans and franks out like shrapnel.

But in the days before helicopters, heating a C ration was more work. Plafker treated it like a ritual and got very good at it. First he'd take a few small rocks and arrange them in a circle, creating a mini fire ring. He'd tear up the heavy cardboard container that the ration came in, as well as the lighter boxes that held some of the individual items. He'd set it all up in a pile in the middle of the ring and add some of the toilet paper as kindling. Then he'd light it with the matches provided. If done just right, the fire would last just long enough to heat the main dish and a cup of coffee.

Plafker teamed with Miller a lot and came to appreciate the older man's skill at working in the backcountry. But Miller was one of the Spartan geologists. He didn't really care about food. To him, a meal was just sustenance—all you needed was enough to keep you going so you could do your job. It didn't matter what it was as long as it filled you up.

What Miller cared most about was getting the fieldwork done right. He could be a taskmaster, especially if the weather was good. He would insist that they stay in the field for as long as possible rather than taking a day or two off back in Cordova. During one string of blue-sky weather, he and Plafker worked a month straight. And this being southern Alaska, where in summer the sun rose early in the morning and didn't set until 10 or 11 at night, the days could be long ones. Plafker, who was in excellent shape, found the work exhausting, and not just physically. Keeping track of all that they were observing, recording and collecting, and doing it day after day, was mentally tiring. But Miller seemed to relish it.

Miller was no daredevil, but he would go up or down any creek necessary to get the needed information about the geology of the landscape so that his maps were accurate. Plafker went up and down many creeks with him, and it was usually an adventure, with all their equipment and supplies loaded onto an inflatable boat. If they were headed downstream and the rapids didn't seem too bad, they'd jump in and paddle themselves. But if they knew the river was rough, or if they were headed upstream, they would attach lines fore and aft and guide the boat from the shore. It was a tried-and-true way of doing things, but things could go wrong. The boat might get tossed in rapids, or a line might break. Miller, undaunted, would press on.

Over three years, Plafker learned a lot—about working in the backcountry, and about geology. He'd spend the summers in

Alaska, gathering data, and the rest of the year back home in the Bay Area with Ruth and Linda. At the office in Menlo Park, he would be busy writing reports, making maps and planning fieldwork for the next year. There was nothing more satisfying, he thought, than finishing a map of an area he'd surveyed, having filled it in with different colors keyed to the types of rocks he'd found, then looking at it and realizing that, geologically speaking, he knew everything about the area that it was possible to know, because he had walked the land himself.

He looked forward to returning to Alaska again in the summer of 1956. But Ruth had become pregnant again and gave birth to a boy, Gary, late in 1955. Maybe, Plafker thought, this wasn't the ideal job after all. With two young children he had to consider finding a job that paid better than his government salary. With his developing expertise, he knew just the place to look: an oil company.

In the 1950s, Chevron, one of the successor companies that had followed the breakup of Standard Oil four decades before, was aggressively looking for oil throughout Central and South America. The company was hoping to replicate the success it had had in Venezuela, where it had discovered the huge Boscán oil field near Maracaibo right after World War II. They had a need for experienced geologists like Plafker, who was quickly hired. The pay was better, and, because he was working for a company and overseas, there were perks like bonuses and housing allowances.

Plafker had first been sent to Guatemala. On the surface at least, the country at the time was largely peaceful—a military junta had deposed the left-leaning government in 1954 in a coup engineered by the US Central Intelligence Agency. Guatemala's long and brutal civil war, with its death squads and other means of government repression, wouldn't begin until 1960. George and Ruth had a large comfortable house in Guatemala City for their growing family—their third child, Jordan, was born while they were there—and a cook, a gardener and a couple of housekeepers to help.

The fieldwork in Guatemala wasn't all that different from what Plafker had done in Alaska, the main change being the surroundings—tropical jungle versus subarctic forest. Plafker worked with mostly Guatemalan crews, and between dealing with them and taking Spanish lessons when he was back from the field he quickly mastered the language.

The main problem, it seemed, was that Guatemala appeared to have precious little oil. In three years of mapping, Plafker never found much in the way of promising formations. So in 1959 he was shifted to Bolivia, where, although the country's petroleum industry had been nationalized for years, foreign companies had recently been allowed once again to prospect for oil. The family moved to Cochabamba, a lovely old city high in the Andes famous for having the best corn beer, or *chicha,* in Bolivia. It was a good base of operations for exploring for oil-bearing formations in the llanos, the large lowland basin to the east.

It was out in these lowlands that Plafker ran into some trouble. He had been working in the office in Cochabamba, but his boss wanted to see what field camps were like. So they took a helicopter to the town of Todos Santos, not far from where a field team was working. The helicopter dropped them off and then left to ferry some supplies to the men at their field camp, but upon returning the pilot clipped a guy wire on a bridge. The chopper crashed and burned, killing the pilot. It was a horrible turn of events, and they had had to call to Cochabamba and tell Chevron higher-ups what had happened and request another helicopter to come and get them out. The helicopter couldn't come until the next day, so Plafker and his boss went to a bar-restaurant to have a drink and get something to eat.

They were minding their own business—while knocking back a few beers to ease the pain of a long, terrible day—when a very drunk bar patron made an offensive remark about gringos, clearly aimed at them. Plafker, who was in no mood to put up with insults and who by then spoke impeccable Spanish, told the man in no uncertain terms to shut up. The man didn't, and punches were thrown. Plafker knocked his antagonist out, but someone called the police. The next thing he knew, Plafker was in a bare-bones jail cell. A short while later the police threw another man in with him—his victim from the bar. The man was now conscious, but mumbling in Spanish so incoherently that Plafker couldn't make out what he was saying. Fortunately he was so drunk he had no idea that he was now sharing the cell with the man who had knocked him out.

Plafker spent the night on the floor of the jail cell—it lacked basic amenities like a mattress—next to the man. The next day, Chevron hastily added a passenger, a lawyer familiar with the justice system in central Bolivia, to the helicopter flight. After some cajoling conversation from the lawyer, and an exchange of pesos, Plafker was set free.

Aside from that night in jail, Plafker's experience in Bolivia had been a good one. He liked Cochabamba and enjoyed the fieldwork down in the lowlands. He still hadn't found much oil, but he'd learned a lot more geology. He was finding that his scientific education never really ended—there were always new things to see and discover. And one thing he discovered in Bolivia would help shape his thinking years later.

In school, Plafker had learned, like everyone else, about the similarities between the coasts of South America and Africa, about Alfred Wegener and the theory of continental drift and about the more general issue of how the mountains and continents formed and changed. After years when the idea that the continents moved had been out of favor, especially among American scientists, there was now a renewed debate about the subject. Plafker wasn't aware of all the details of the new evidence that was being uncovered—he was a field geologist, not a geophysicist—but he knew that now a growing number of scientists could be considered mobilists rather than stabilists.

He'd always been convinced, from those clam broth and beer–

fueled field trips with A. C. Hawkins, that it was important to see things firsthand. His work in Alaska and in Guatemala had reinforced the view that there was no substitute for being on the ground, in the field. In the field in Bolivia he'd seen things that had made him think more about the continents. And he began to believe that Wegener had been on the right track.

While examining sedimentary rocks in the eastern Andes foothills, Plafker had come across dropstones. These were rocks, sometimes little more than pebbles, sometimes the size of boulders, that were embedded in layers of much finer sediments. It was as if when the region was covered in water and the sediments were building up, these rocks had been dropped into them. In fact, geologists knew that that is one way dropstones occur—they are picked up by glaciers, become frozen in the ice and carried along and then, as the front of the glacier calves off, form icebergs, eventually dropping out as the ice melts.

The problem, Plafker realized, was where had the glaciers been that had done this? They had to have come from the east, but to the east was only more lowland in Bolivia and Brazil. But if South America and Africa had once been one—if the South Atlantic Ocean had yet to form—perhaps the glaciers could have come from there, or even from what is now Antarctica. The dropstones seemed like obvious evidence that the continents moved, that the earth's surface was not static. Plafker was well on his way to becoming a mobilist.

He would try to talk about what he was seeing with the other geologists he was working with. But they were largely uninterested. Because they worked for an oil company, they had a focused approach to geology. They wanted to know if a certain rock formation had the potential to hold oil or gas, and, if so, whether there was enough so that getting it out of the ground would be a profitable venture. Beyond that, Plafker came to understand, they didn't really dwell on whether certain rocks had once been somewhere else, or how they had gotten to where they were now.

Plafker, on the other hand, was intrigued by these sorts of

things. Perhaps oil work was not for him, he thought. But it paid better than the Geological Survey.

All of this was on his mind in 1961 when events conspired to alter his career once again.

Chevron called an all-hands meeting in Bogotá, Colombia. The company was going to pull out of the region, Plafker and others were told. We have all this oil in the Gulf of Mexico that we can get for $3.00 a barrel, they said. Saudi Arabia has most of the rest, which they can get for $1.50 a barrel. How can we compete with that in places like South America?

They offered Plafker a couple of options. He could transfer to Chevron's office in Oklahoma City, where he would work on the subsurface geology of the many oil and gas wells they had there. The idea of moving to the mountainless middle of the United States and working in an oil patch was less than appealing. Or, Chevron told him, you could work for our partner, Shell, which is exploring for oil in Libya, in North Africa. That sounded much more interesting to Plafker, although working for Shell, a Dutch company, would mean less in pay and perks.

He was mulling over the idea of moving to Libya when a letter arrived from Don Eberlein, his former boss at the Geological Survey's Alaska branch. It contained bad news. Don Miller, Plafker's old mentor, had died in Alaska in August. He and an assistant, Bob MacColl, had drowned in the Kiagna River, on the north side of the Chugach Mountains. Miller's luck, it seemed, had run out.

The letter included a few details, which Plafker fleshed out later. Miller and MacColl had been working in the area and had chosen to go down the Kiagna, a north-flowing glacier-fed river about 150 miles northeast of Cordova, mapping rock exposures along the way. Miller was traveling the same way he had with Plafker. The two men had loaded supplies and equipment in an inflatable boat, tied ropes to it and guided it down the river. But they also had access to a helicopter to move them from one locale to another if needed.

After a reconnaissance flight in the copter in early August,

Miller, ever cautious, had decided that it would be safe to travel only through the lower end of one of the deep canyons along the Kiagna, as the river was running too high elsewhere. The helicopter had dropped the two men at the canyon's lower end, landing on a gravel bar in the middle of the river.

All went well for the first day or two. A bush pilot who was to fly over each day to check on them saw them working in the canyon on August 5. But the next day the weather turned bad, and when the pilot flew over the river late in the day he saw the bodies of the two men on a bar, below the mouth of the canyon. Whether Miller had miscalculated the potential flooding danger or there had been some kind of accident, it was impossible to tell.

It was sad news for Plafker, who took several moments to let the information sink in. But then he read on. Eberlein had a question for him: We've just lost one of our best field geologists, a man who knew southern Alaska like the back of his hand. You worked with him and know the area just about as well as he knew it. How would you like to come back to the Survey, and to Alaska?

7

BEFORE THE STORM

If a single quotation best reflects the Alaska of the mid-1960s, it might be one that actually was spoken years later, by Walter J. Hickel, a towering figure in the modern history of what natives called *Alyeska,* the Great Land. Hickel was one of the men who made Alaska, but Alaska also made him. Raised in Kansas, he'd come north in the early 1940s and gotten into the construction business. That led to real estate development, and before too long Wally Hickel, as he was universally known, was a major force in Anchorage's growth, building shopping centers, housing and hotels. In the 1950s he also became enmeshed in Republican politics, and he was one of the leaders who helped ease the way to statehood in 1959. Hickel was the state's second governor (defeating Valdez's own Bill Egan, who was seeking a second term) and its eighth as well, having been elected again in 1990 after taking time out to serve as Richard Nixon's interior secretary, among other things.

The words in question were spoken by Hickel during his second gubernatorial term, in 1992. At the time, state fish and game officials had approved a program to bring Alaska's gray wolf population under control. The animal was for many Alaskans a noble, if unofficial, symbol of the state. But the gray wolf was also a major predator of other wildlife. So the state had decided to allow some to be shot—by fish and game agents from airplanes in some areas and by private hunters in others. Their rationale was that thinning the ranks of the predators would increase the numbers of caribou, moose, bears and other game, creating an Eden for hunters and, for tourists, a spectacle that state officials rather extravagantly claimed would rival that of the Serengeti in East Africa. The plan had been approved despite objections from environmentalists and some wildlife biologists, who argued that the wolf was part of a complex ecosystem and that the state shouldn't play God with it. Altering the predator–prey balance, they pointed out, could paradoxically lead to fewer big game animals, as populations might first increase so much that competition for food would become fierce, and starvation could ensue.

Hickel had been a supporter of the plan to cull the wolf population—he was a booster of almost everything that would encourage tourism and other business activity in the state—and, when he was asked about it by a reporter, had responded with a memorable line.

"We can't just let nature run wild," he said.

Modern Alaska, it could be argued, lived by that concept. The state had been, and still was, defined by nature. Most of its territory was wilderness, tens of millions of acres of land that far outmatched, in breadth and often in ruggedness and grandeur, anything in the Lower 48. But Alaska had grown, and would continue to grow, because Alaskans had brought nature under control when and where they could.

By 1964, the state's population was about 250,000, including 40,000 natives. Far from being a land of hard-drinking sourdoughs and grizzled mountain men living in cabins made from rough-hewn logs—though those stereotypical characters existed—Alaska was well on its way to becoming much like the rest of America, urbanized and suburbanized. Fully half of the population lived in or around Anchorage and Fairbanks, the state's biggest cities.

Of the two, Anchorage was bigger and had come the furthest. Like Valdez, Anchorage was first a tent city, although a more organized one, having been deliberately established in 1914 at the head of Cook Inlet, where Knik Arm branches to the northeast and Turnagain Arm to the southeast. Several thousand workers who had been enlisted to lay the railroad from Seward to Fairbanks made their home at an encampment along Ship Creek, just north of, and down the hill from, what was to become downtown. As in Valdez, the tents had soon been replaced by buildings, and Anchorage became a hub for the new rail line, which after wartime delays was completed in the 1920s.

Since then Anchorage had expanded, venturing forth into the wild, as it were, and taming it. The rail line and an ever-growing military presence—Anchorage became the headquarters of the Alaska Defense Command at the start of World War II—helped feed the city's growth. By 1964, thanks to the efforts of Hickel and others, Anchorage was booming. Its downtown was bustling with shops, restaurants and commercial buildings. A new five-story J. C. Penney department store, a gleaming, nearly one-hundred-thousand-square-foot retail showcase, had opened on Fifth Avenue, just off the popular Fourth Avenue entertainment district with its bars, arcades and movie house. Anchorage was awash in government offices as well. Although the capital was in far-off Juneau in the Panhandle, as the center of Alaska's population Anchorage by necessity handled much state and federal business.

In Anchorage's earlier days, planes had landed on a strip of parkland just a few blocks south of downtown. In 1951 a real airport was built southwest of downtown, and it was now buzzing with jetliners coming from Seattle and other American cities and with European flights stopping to refuel on their way to Asia. Anchorage was also starting to become a hub for a new way to ship cargo across the Pacific, by air.

Hemmed in by water to the west and the slopes of the Chugach to the east, the city had spilled out any way it could: to the south around the airport and along the Seward Highway, and to the north, past the air force and army bases and up the eastern shores of Knik Arm toward the broad valley of the Matanuska River.

The valley had seen its own development, beginning in the nineteenth century with the discovery of abundant coal deposits north of what soon became the settlement of Palmer, about forty miles from Anchorage. The area had always attracted homesteaders interested in farming, lured by the rich valley soil, but in 1935, as part of a New Deal program to help hard-hit farmers in the Midwest, a horde of newcomers arrived. The federal government moved more than two hundred families from Wisconsin, Michigan and Minnesota—states that were chosen in part because their climate was much like Alaska's—to the valley, where they drew lots for forty-acre plots. The Matanuska Colony, as it was called, wasn't a complete success, as the short growing season hampered agriculture, and over half of the original colonists eventually left. But the influx of people and money—the project is estimated to have cost $5 million—did spur the development of Palmer. Far from being an outpost in the wilderness, Anchorage was the center of what was practically becoming a sprawling metropolitan area.

The state's growth wasn't limited to that region, however. Fairbanks had also vastly expanded beyond its origins as a dusty trading post established at the turn of the twentieth century to support gold prospectors working claims to the north. Early on, Fairbanks had briefly become a boomtown—one four-block stretch had no fewer than thirty-three saloons—but gold fever had petered out around the time of World War I. The road from Valdez, and the railroad from Anchorage and points south, enabled Fairbanks to continue thriving as the commercial capital of interior Alaska. Like Anchorage, it benefited from the presence of the military during World War II and especially when the Cold War thrust Alaska into prominence as a first line of defense against Soviet attack. The Pentagon spent hundreds of millions of dollars on airfields and other more exotic installations built for programs like White Alice, a sprawling communications network of microwave links and large antennas that seemed to rise up out of nowhere on the landscape, and the Distant Early Warning, or DEW, Line, an array of coastal radar stations built to detect a Soviet nuclear first strike.

Civilization was thriving elsewhere as well. South of Anchorage, Seward and Whittier benefited from being on the rail line.

In the Panhandle, Juneau thrived as the state capital, and on the Copper River delta, at the eastern edge of Prince William Sound, Cordova became a center of the salmon fishery.

Since World War II, a statehood movement had been growing among Alaskans who chafed at being ruled largely by politicians and bureaucrats thousands of miles away in Washington, D.C. Alaska was such a different place from the rest of the United States, they argued, that it especially needed to be governed by Alaskans. Despite (or perhaps because of) strong opposition in Washington, the movement grew in the mid-1950s, reaching a milestone with the writing of a constitution that was ratified in 1956. Two years later, Congress finally approved statehood, and Alaska entered the union as the forty-ninth state, on January 3, 1959.

Within a few years, much of the bloom had come off the rose of statehood, in large part because of the fiscal shocks it generated. (Indeed, concerns about how Alaska could afford to govern itself had been behind much of the opposition in Washington.) The state was suddenly on the hook for services that the federal government had previously paid for, like highway construction and maintenance, and also had to build the infrastructure of state government. Statehood had also done little to settle the land claims of Alaska's natives, an issue that lingered until 1971.

But statehood had given Alaskans more freedom to pursue their economic goals. The state's fisheries, buoyed by new regulations, began to take off. Gold and other mineral resources still supplied a steady stream of revenue. The long battle to become a state had raised Alaska's profile in the rest of the United States; a tourism industry grew as vacationers started coming north to see what the fuss was about. The military was still a strong economic engine. And Alaska was beginning to get a glimpse of a future that would be drenched in petroleum: it received its first oil revenues, of some $3 million, in 1959, from oil and gas fields discovered on the Kenai Peninsula. In 1968, the first oil reserves were found at Prudhoe Bay on the North Slope.

So five years along, Alaskans could feel confident about the direction of their state. They'd made the sometimes-rough transition to statehood and survived the financial squeeze it created. There were still limitless amounts of wilderness—one of the provisions of statehood was that Alaska got to choose more than one hundred million acres of federal land for its own—but Alaskans were making inroads with their own special brand of civilization. As Ernest Gruening, one of Alaska's first senators, put it in a long encomium to the state published in *National Geographic* in 1959, Alaska's people were "enterprising, vigorous, warmhearted, modern. In the larger communities they shop in supermarkets and neon-lighted drugstores, read everything from comics to classics, and watch television."

Alaskans were living, and enjoying, a distinctly far-northern version of the American Dream. It was one that relied on not letting nature run wild.

Nature, though, was about to strike back.

With little more than a month left in the school year, Kris Madsen was growing restless. The isolation of Chenega, plus the usual cabin fever after a long winter—the island was not as bitingly cold or snowy as some parts of Alaska, but it could be dark and dreary—had proved to be a bit much. Perhaps sensing that she needed a diversion, however brief, a bush pilot who often stopped at the island had taken her up in his floatplane one day for a sightseeing flight. He was just having pity on me, Madsen thought, but she appreciated the chance to see places that her students had talked about, like the cannery at Port Nellie Juan and the awe-inspiring mountains and glaciers that framed Prince William Sound.

The teaching had been going well enough, despite a lack of proper books. She'd discovered, much to her dismay, that there were no primers, or beginning reading books, for her youngest students and that most of the books for older students were hopelessly out of date. But Madsen had made do, and also had learned how to handle a room full of students of differing ages. She genuinely liked most of her charges, though they would sometimes bring disputes and other problems from home to the classroom. And she'd done more than teach. Although she had no training and very little in the way of supplies, she had been the village's de facto nurse, dealing with the cuts and scrapes of everyday life. She'd cared so well for eleven-year-old Timmy Selanoff when he sliced his thumb open down at the beach one day that the real nurse, who visited Chenega irregularly for inoculations and other routine medical care, had been impressed. Madsen also had dealt with unexpected events, especially the news that came over the radio on a Friday in late November that President John F. Kennedy had been assassinated in far-off Texas. Chenegans revered the president; many had his portrait on the walls of their homes. Madsen had sent her students home early and established an informal period of mourning, lowering the flag to half-staff and canceling classes that she had scheduled for the next day.

But she knew that a year in Chenega, after a year in Old Harbor, would be enough. She was thinking about what she would do next. Perhaps she'd get a job in Anchorage or even Fairbanks, farther north. But the state educational bureaucracy was stifling. She had endless reports to fill out and send off via the mail boat, and endless testing that the state wanted her to inflict on her students. One request was particularly galling: an IQ test with questions that she felt were laughably inappropriate for young native people whose world was a small isolated community like Chenega. One question featured a photograph of a filling station, something many of her students had never seen. Madsen had put big X's across the answer sheets and sent them back to the bureaucrats in Anchorage. So perhaps, she sometimes thought, when the school year was over she'd end her Alaska adventure and return to sunny and warm California for a while before heading somewhere else.

She would figure it all out eventually. On this day—overcast but not too cold, with temperatures in the thirties—she had teaching to do. Before her students came up the steps that morning, she raised the flag on the pole outside and readied her classroom for the day, arranging the desks and chairs and making sure that whatever books and supplies she would need were set out. She checked the surplus food in the storeroom for her students who had not brought anything for lunch or who, later in the day, might need a little something to tide them over. Then she did what she always did: she took up a piece of chalk and, in her plain teacherly hand, wrote the date on the blackboard.

Today is Friday
March 27, 1964

It was Good Friday by the American church calendar. But Russian Orthodox Easter was not until May 3, more than a month away. So while school might have been out in Anchorage and other places, it was business as usual at the schoolhouse on the hill. Madsen needed to teach every day she could—that's why she taught some Saturdays—because the state required 180 days of classes. No one knew precisely when in the spring the salmon would start running, but it would be sometime in May, and once they did her students and their families would be on their way to Port Nellie Juan and the school year would come crashing to an end.

She'd learned that one of the tricks to teaching different grade levels at the same time was to assign an in-class project that everyone could work on at his or her own pace and that older students could help the younger ones with. Then she could work with each age group on academic subjects as needed throughout the day. That day, the children worked on paper Easter eggs to hang on the wall. She'd let the youngest go shortly after lunch, and the older students had left at about three. She'd spent the rest of the afternoon tidying up the classroom for the big event that evening. For this Friday night was movie night in Chenega.

This was another of Madsen's responsibilities that no one had told her about. Chenegans' only contact with the world was through the radio. Most villagers had one, powered by batteries, and there was a two-way radio at the school. Movie night was a way to bring the outside world to Chenega—filtered through the lens of Hollywood, of course. It was also one of the few nonreligious social activities that involved practically the whole village. Movies were shown on Friday nights, although not every week. The schoolhouse became the movie house, with a 16mm projector that Madsen had learned to operate.

That evening's feature was *House on Haunted Hill*, a grade B horror flick that Madsen had ordered through the mail, as usual, and that had arrived on the mail boat. The film, released to little acclaim in 1959, starred that master of horror, Vincent Price, as an eccentric millionaire who invites five people to be locked up in a Gothic mansion overlooking Los Angeles, with a promise of $10,000 to whoever can make it through the night. Needless to say, murder and mayhem ensue, aided and abetted by the millionaire's fourth wife, played by Carol Ohmart, who a few years before, like countless young starlets, had been billed as "the next Marilyn Monroe."

Madsen, with the help of her friend Norman Selanoff, had moved the desks in the classroom off to one side and set up chairs and a screen. She had threaded the first reel of the film. There was no set showtime—the movie would start whenever it seemed that most of the villagers made it up the steps after dinner, perhaps around 6:30. With some time to kill before making dinner herself, Madsen and her friend decided to fetch some water from the pond above the school. Normally the water flowed down a pipe from a small dam on the pond. But there'd been a week of abnormally cold weather recently, and the pipe was frozen, restricting the flow to a trickle. So they would have to fetch the water in buckets.

Down in the village, the day had been an ordinary one, overcast but not too dark, and not too cold. Some of the men had spread fresh gravel around the outside of the church, a customary way of sprucing it up during Lent. Others had gone seal hunting: Mickey and Nick Eleshansky had taken their boat, the *Shamrock*, over to Prince of Wales Passage, about a dozen miles from the village, and George Borodkin and Mark Selanoff were in their boat, the *Marpet,* near Icy Bay, about half that distance away. Nick Kompkoff had just painted his eighteen-foot skiff the day before and was thinking about taking it on a quick hunting trip, but the paint was not quite dry, so he decided to stay home. Later in the afternoon he wandered over to the Smokehouse to shoot some pool.

Mary Kompkoff, Nick's wife, had done some cleaning earlier in the day and now had a pot of chili cooking on the family's oil-fired stove. With Nick out of the house, she decided to do a little visiting herself before supper and the movie. She turned the stove down as low as it could go and went over to see her sister, Dora Jackson. Dora, who at twenty-eight was a year older than Mary, was separated from her husband, Nick, and was raising her three children—Cindy, eight, Dan, four, and one-year-old Arvella Jane—by herself. Mary found Dora in a bit of a melancholy mood, ironing her curtains, so she didn't stay long. But rather than going home she decided to go see her younger sister, Shirley Totemoff, who was sitting around her house, waiting to go up the hill for the movie.

Other villagers were passing the time before the movie by indulging in a steam bath. For those homes that had one, the bath was a small structure attached to the smokehouse. The stove in the smokehouse was used to heat river rocks, which were then moved, using wooden paddles, to the steam bath and put in preheated water that flashed into steam. The effect was instantly invigorating.

After school, some of Madsen's students had gone down to the beach to play, taking advantage of the low tide. Nick Kompkoff Jr., Nick and Mary's nine-year-old son, his eleven-year-old brother, Mark, and a friend, Jerry Lee Selanoff, ten, went along the beach to the right of the village. Nick and Mark's sisters—ten-year-old Julia Ann, Carol Ann, four, and Norma Jean, three—were playing on and under the dock. And a few of the Selanoff kids—Timmy and his brothers Kenny, thirteen, George, six, and Billy (known by his nickname, Buttons), five, and four-year-old sister, Jeanne (known as Gula)—were down there too. Kenny had wanted to chase and throw stones at birds, and the rest of the clan followed along. The two youngest, Buttons and Gula, headed off in the same direction as Nick Jr., Mark and Jerry Lee, although the three older children soon grew tired and thirsty and headed up the bulkhead steps to Aunt Shirley's for a drink. Kenny and the others went the opposite way along the beach. Timmy, ever the competitor, wanted to beat his older brother at this game, so he ran after the birds a little bit faster, even though the pockets of his black jacket were weighed down with "ammo"—stones he'd picked up. Soon he was by himself a long way down the beach, far away from the village and from the other boys.

In her home along the bulkhead, Avis Kompkoff was getting ready to take a bath. Earlier she had gone over to the house of her aunt, Margaret Borodkin. The two had gone out in a skiff the day before looking for sea cucumbers and were surprised to find that the whole bay was full of them. Margaret liked to eat sea cucumbers, but Avis didn't—as she'd say, she would never eat something that bumpy.

Today, however, she wanted to see about taking a steam bath at Margaret's house, as her house didn't have one. But her aunt had told her that there wasn't enough heated water, so Avis had gone back home with the idea of taking a regular tub bath. She had taken a five-gallon tin of water and put it on the stove.

Her daughter Jo Ann, now three and a half, was with her, as were two-and-a-half-year-old Joey and the baby, Lloyd, who was not quite six months old. Her husband, Joe, was down at the beach. A cousin, Richard Kompkoff, had been visiting at the house but was getting ready to walk to his home, over near the church.

Those days Jo Ann slept at the house of her grandmother and grandfather, the Evanoffs, which was also in the center of the village. Avis had a thought—Richard could take Jo Ann there on his way home. She asked the girl if she wanted to go home to be with her grandmother. Jo Ann said yes, so Avis asked her cousin if he would drop her there as soon as she got her dressed. Then she turned to Joey and asked him if he wanted to go with his sister. But before Joey could say yes, she had made up her mind—he was staying.

In a little while Richard ambled off with Jo Ann in tow, wearing a favorite pink T-shirt. Later, Avis would recall that the reason she sent Jo Ann to the Evanoffs' house that day was that she'd heard a voice in her head. "Send her home," it said. The same voice had told her to keep Joey with her.

By the standards of mid-1960s America, when it came to entertainment, Valdez was a bit behind the times. Perhaps that was to be expected of a community that was so far from other outposts of civilization, at the end of a long fjord and an even longer road. Not that the citizens of Valdez seemed to mind. There was a movie theater, at the Eagles Hall, with screenings four days a week, on Saturdays, Sundays, Tuesdays and Thursdays. Some townspeople were regulars—a ticket was less than a dollar—and treated the place more like a community center than a theater, getting up and gossiping with neighbors and friends while the projectionist changed film reels. But there was a real community hall too (it doubled as a museum during the summer months), where the Teen Club held dances on Friday nights and the Sour Docees, a square dance club, held them on Saturday nights. The two would join forces once or twice a year to cut firewood to keep the place heated.

But Valdez had no radio stations and certainly no television service. While the rest of the country was tuned in to *The Adventures of Ozzie and Harriet* or *Walt Disney's Wonderful World of Color,* Valdez was in the dark. TV didn't come to the area until the early 1970s, and even then shows were at least a week old, having been taped in Seattle and shipped north.

The people of Valdez found other ways to occupy their time. With its long winters, the town, like most communities in Alaska, was crazy about basketball. The high school had a perennially good team, taught by a popular coach and teacher, James Growden, and the traditional season-opening student-alumni game was front-page news. There was an adult league as well, with teams sponsored by local businesses, including some of the bars. Valdez also had clubs of all kinds—garden, chess and women's among them—and civic associations like the Fraternal Order of Eagles and the Loyal Order of Moose. Residents were also voracious readers, apparently—in 1959 the state librarian noted that, per capita, Valdez had more books checked out of its small library than any other city in Alaska.

The Fourth of July was celebrated with a small parade, but the real fun was saved for Gold Rush Days, a multiday festival later in the month that was first dreamed up by town boosters in 1962. The activities included a pageant to select a queen (in 1963 one contestant listed "swimming and clothes" as her interests), a variety and talent show, a parade and, topping it off, a grand Coronation Ball, for which everyone dressed up as sourdoughs. For weeks before the event, many of the men in town took to growing beards to better look the part. "It is a real inspiration to see the number of beards being grown," a local observer wrote in 1963. "And some of them aren't too bad looking, either."

But with relatively little in the way of formal entertainment, especially during the cold and snowy months, the people of Valdez were suckers for anything that broke the routine. If there were reports that a black bear and her cubs had been sighted up a tree on the Robe River off the Richardson Highway, townspeople would hop into their cars and go have a look.

Nothing, however, broke the routine more than the arrival down at the waterfront of one of the Alaska Steamship Company's converted Liberty ships on a regular cargo run from Seattle. In its holds was just about everything Valdez needed: food, fresh and canned, for its grocery stores and restaurants; dry goods, furniture and clothing for other shops; automotive parts and tires; appliances, construction equipment and supplies of all kinds. Other cargo would be off-loaded and stored in one of the warehouses, eventually to be trucked to towns along the Richardson Highway all the way to Fairbanks.

When a ship came in, a crowd would gather. Men would be there to work, having been hired by the Valdez Dock Company as longshoremen to unload the cargo. For a large ship, about a dozen men were needed to work on board, including eight who would be down in the holds. Dockside, a gang of six to ten would handle the cargo as it came off the ship. Picking up a shift was a way for some of Valdez's breadwinners to supplement their income, especially outside the summer season. The pay was good: in the mid-1950s the going wage was more than three dollars an hour, and nearly twice that if the crew was unloading ammunition.

Others came just to watch. Some were women and children who were there simply because their husbands and fathers were longshoring. But for parents with young fidgety children, or for older kids, the unloading of a cargo ship was a spectacle. This was long before security concerns made waterfronts off-limits to the public; at Valdez, the curious could walk—or drive, for that matter—onto the dock, and as long as they didn't get in anyone's way could stay as long as they wanted. They might even be able to go on board the ship, if the crew were friendly and the captain approved.

This was also in the days before widespread containerization took what little romance there was out of the cargo trade by hiding all the goods in twenty- or forty-foot corrugated steel boxes. Most of the cargo that came to Valdez was stacked on pallets that had to be lifted out of one of the ship's five holds by a derrick mounted on the deck, swung over the side and lowered to the dock. Those watching could get a general idea of what was being unloaded, and with all the booms, cables and ropes, with forklift trucks moving cargo into warehouses and trucks arriving to haul some of it away, there was enough activity to satisfy onlookers for hours.

On March 27, 1964, it was the SS *Chena* that had steamed into port, shortly after 4 p.m. For the young people the day had already been special, because with the Good Friday holiday there had been no school. The ship's arrival just added to the good mood.

Built in 1942 in Portland, Oregon, the *Chena,* like all the more than 2,700 Liberty ships made for the war effort, was precisely 441½ feet long from stem to stern and 57 feet wide at midships. And like many ships among the hastily built fleet, it had been made with brittle steel. Early in its service, that steel had failed spectacularly. Then called the *Chief Washakie,* the ship had been off Unalaska Island in the Aleutians one night in December 1943 when the stress and strain of rolling waves caused its hull to crack. The ship had barely made it back to Dutch Harbor for repairs. The steamship company had picked it up as surplus in the 1950s and renamed it for the Chena River, which runs through Fairbanks.

On board the *Chena* that day was a crew of thirty-nine, led by the ship's captain, Merrill D. Stewart. Of all the ships that called at Valdez, the *Chena* held a special attraction for some of the town's young people. Crew members were known to throw candy down to those on the dock. And the ship's cook would hand out oranges or other fresh fruit to those in the know.

Danny Kendall hadn't been one of those in the know. Danny's family had arrived in Alaska in 1957, soon after his father, Bill, had been discharged from the navy. Bill Kendall had been stationed in the San Francisco Bay Area, but he and his wife had decided to seek their fortune elsewhere. They had loaded the five kids into the car—Danny was just a toddler then—and had driven up the coast, looking for work along the way. They hadn't had much luck until they got to Seattle, where they found the newspapers full of want ads for jobs in Alaska. So they kept driving north.

Bill Kendall had worked for the Alaska highway department, first in Anchorage, then in Juneau and finally in Valdez. There, he'd eventually gotten a job with the city administration. But by 1964 it seemed as if his days in town were numbered. He'd proposed that Valdez reopen a World War I–era hydropower plant at Solomon Gulch on the Lowe River. The proposal had divided the town and was opposed by some of the town fathers—people like Owen Meals, who owned the current diesel generating plant and saw no reason to switch. The matter had been put to a vote, and Kendall's proposal had lost. There had been bad feelings on all sides.

Danny had heard about the *Chena*'s cook from his older brother and some of his brother's friends. So with the *Chena* in town on that Friday afternoon, when he saw two of those older boys—Dennis Cunningham and Stanley Knutsen—heading toward the waterfront, he tagged along. When they got there, the two older boys headed onto the ship. Danny tried to keep up, but it soon became clear that Stanley and Dennis wanted nothing to do with him. Danny got the message; he left the waterfront and walked home, leaving Dennis and Stanley on the dock.

At his family's house outside of town, one of Danny's friends, Gary Minish, was enjoying not having to be in school. The house was at Mile 2 of the Richardson Highway, on seven and a half acres that the family had homesteaded since the early 1950s. Like Danny's father, Gary's father, Frank, had come to Alaska fresh out of the military. During the final months of his service he had been stationed in Valdez, and that was enough for him to fall in love with the place. The economic prospects back home in South Dakota were dismal, postwar Alaska had a lot of new construction going on and Frank knew carpentry. And then there was the homesteading law: all you had to do was stake a claim and build a structure on the land and it would be yours eventually. As soon as he was discharged, he'd told his wife that they were going to load the truck and head north. Gary was a year old at the time.

Like so many Valdez men, Frank Minish picked up work when he could down at the docks. This day he was due to work the *Chena*, and since the ship hadn't been scheduled to arrive until late afternoon, he'd spent much of the day sleeping. But when he got up in late afternoon, he wasn't feeling particularly rushed. I'm always the first one down there, he told the family. Let everybody else go first today. I'm going to have some breakfast.

Not so the neighbors. The Stuarts—Earl (whom everyone knew as Smokey), his wife, Sammie, and their three children, Janice, Larry and Debra—had packed themselves into the family car and headed down to the dock a short while before. Smokey was working a shift—he was employed at the highway department and could use the extra money—and Sammie and the kids were just going to watch. It was something to do, and besides, the family had only one car.

In her bedroom in the family home on Broadway Avenue, Dorothy Moore was busy ironing the dress she was going to wear on Easter, two days away. Her father, Sid, was at work, helping to run Owen Meals's power plant. It was a short commute, as the plant was just across the street.

Dorothy, nineteen, the oldest of six children, was back in town for the holiday from Anchorage, where she was studying at Alaska Methodist College. She had lived in Valdez practically all her life, but truth be told, hadn't really cared all that much for it. It was a small town, she thought, with provincial attitudes, and a place where everybody knew everybody else—and that wasn't always such a good thing. Growing up, she'd had to endure her share of unfounded gossip. One particularly nasty, and untrue, rumor was that Dorothy, with her dark hair, was a child of her mother's first husband, since all the other Moore children had red or blond hair.

Dorothy had been a toddler when her parents had moved to Valdez in 1949. By the standards of many Alaskans, the Moores were old-timers. Her grandparents had come north from the Lower 48 shortly after the First World War and had settled along the lower Yukon River—the Kuskokwim area—for several decades before coming to Valdez. Her grandmother hadn't cared all that much for Valdez either, finding coastal Alaska very different from where she'd lived before. But she'd gotten involved in town life, and the family had taken root here.

Gloria Day had taken root here too. She had come to the city at age twenty-three, right after the war, escaping from a bad marriage back in her hometown of Columbus, Ohio. She'd told her parents that she needed a break, so they had agreed to look after her two young children while she traveled. She'd made it to Anchorage, but it wasn't far enough for her, so she'd taken a bus to Valdez and within two days was working at Gilson's grocery store. Two weeks after that, a man walked into Gilson's and announced that he was in need of a bookkeeper. She had done bookkeeping back in Ohio, so he hired her. She ended up marrying his son, Walter Day.

In the years since, Walt Day had held a lot of different jobs, including working for the highway department. Gloria had held several different jobs as well, a stint as postmaster among them. In the early 1960s, seeing an opening for a business in town, the couple had bought a small printing press and started putting out the *Valdez News*, once a week.

She and Walter had spent much of Thursday night as they usually did, printing the paper, and most of Friday morning getting it ready to be mailed. Gloria dropped the papers off at the post office, and they were in most subscribers' mailboxes that afternoon. Back home at her three-story house on McKinley Street, she had taken a bath and put on her pajamas. The couple's daughter, Wanda, who had been out with a friend, had returned as well, but their son Pat was out with the family's red pickup truck. He had told them he was going down to the small boat harbor to pump out their boat, the *Bulldozer*.

The twelve pages of the March 27 edition of the paper contained the usual mix of boring news and lively chatter. The main story was a good example of the former, about a meeting of mining and petroleum engineers near Fairbanks. The speaker had noted that mineral production in Alaska lagged behind that in western Canada, putting the blame for it on federal laws and regulations. It was a viewpoint that would have sat well with many in Valdez, who like other Alaskans had hoped statehood would lessen, not increase, the burdens imposed by Washington. There was a lengthy, and dull, report on an American Legion banquet held up the Richardson Highway in Glennallen, where Governor Egan, proudly identified as a Legionnaire of Valdez Post No. 2, was the main speaker. According to the account, the governor "reviewed his recent journey behind the Iron Curtain on matters of fish control in Alaskan waters."

Like any small-town paper, the *News* reported on developments and activities big or small. Nila Tyler, a high school student, had just returned from a science fair in Anchorage, where she had won a blue ribbon for her project on worms. The Dorcas Club was holding an Easter egg hunt at the elementary school Saturday morning; forty-five dozen eggs were dyed and waiting to be hidden. The Elks and the Cub Scouts had scheduled a pancake breakfast at the Eagles Hall. Saint Francis Xavier Catholic Church was holding a bake sale at Gilson's market, starting at noon, with a drawing for a four-foot floppy-eared stuffed rabbit toting a basket full of candy. There was no shortage of events to satisfy a Valdezan's sweet tooth.

On the bill at the movies the coming week were *Captains Courageous,* with Spencer Tracy and Mickey Rooney, and *The Big Knife,* starring Jack Palance and Shelley Winters. Club Valdez, one of the town's restaurants, advertised an Easter Sunday smorgasbord that included turkey and all the trimmings, as well as ham and baked beans, for three dollars. The Lions Club promoted its spring cleanup campaign. Chaperones were announced for an upcoming teen dance at the high school. The school's chorus was to give a performance of the operetta *Adventures of Tom Sawyer* the following Saturday. "Don't be surprised if you see a portable bush producing music," the announcement read. "It is just our accompanist, Sally Huddleston."

And there, on page 7, was a brief announcement of the upcoming nuptials of Linda Gilbert, daughter of Kenneth Lee Gilbert of Lambertville, Michigan, to Gerald Zook, son of Mrs. William Zook of Valdez. The wedding was to take place on April 4 at 10 a.m. at St. Francis Xavier, with a reception for the young couple following the ceremony at the Switzerland Inn.

Jerry Zook, age twenty-seven, had been born in Wrangell, Alaska. He'd served in the navy, and when he was discharged he'd moved to Valdez to be near his mother and stepfather. A good basketball player, he'd been recruited to play for the alumni that fall in Valdez High's season opener. Somewhere along the line he'd fallen in love.

An employee at the state highway department, he nonetheless was open to taking what extra work he could find, with the prospect of starting a family. On that Friday, eight days before his wedding, Jerry had gone down to the dock to help unload the *Chena.*

8

FAULTS

Howard Ulrich maneuvered his forty-two-foot salmon troller, the *Edrie,* past the spit of land that marked the entrance to Lituya Bay and into the bay proper. It was about 8 p.m. on July 9, 1958, and Ulrich, who had his seven-year-old son, Howard Jr., with him, had decided to put in for the night. He had heard there was good fishing along this wild stretch of southeastern Alaskan coastline, 140 miles north of his home in Sitka. So he had bypassed more familiar trolling spots along the way. Ulrich figured he'd fish first thing in the morning; for now he and the boy, whom everyone called Sonny, would have supper and turn in early.

Lituya Bay forms an indentation in a shoreline that features some of the highest coastal mountains anywhere in the world. The bay is about seven miles long and less than two miles across at its widest, with a small island in the middle. It is ringed by steep slopes that rise up to six thousand feet. From the air the bay looks something like a fish, with a narrow mouth formed by the spit of land and a T-shaped tail at the opposite end, where the slopes create a nearly vertical wall.

Ulrich anchored just inside the entrance, along the south shore in thirty feet of water. There were two other fishing boats in the bay, the *Badger* and the *Sunmore,* both with two-person crews, and a team of Canadian mountaineers camped on the spit after climbing 15,525-foot Mount Fairweather, the highest peak on the coast, about ten miles to the north. As Ulrich trailed off to sleep at about 9 p.m., he heard the sound of an aircraft, a Canadian Coast Guard floatplane that landed, picked the climbers up and departed. Other than that, the night was breezy but quiet.

An hour later, he was awakened by a violent rocking. Ulrich knew instantly that it was an earthquake, and a strong one at that. But he wasn't prepared for what happened about two minutes later.

From down the bay, he heard a deafening crash. The quake had shaken loose an enormous amount of rock and snow from a mountain at the back of the bay. In seconds, by later estimates, fifty million cubic yards of material fell into the water. It created the kind of splash that happens when a rock is dropped into a pond—only in this case the rock weighed some seventy-five million tons and was dropped from a height of three thousand feet.

Ulrich saw the splash as he gazed down the bay. It was enormous, he thought—so far beyond anything he'd seen before that he couldn't begin to gauge its size. But then he saw a wall of water emerging from the lower part of the splash. What had started as an explosion of water was now a mammoth wave. Ulrich guessed it was five hundred feet high. To his horror, it was coming his way.

He woke Sonny, started the engine and tried to pull up the anchor, to reach deeper water and avoid being pounded into the shore. The wave seemed to be losing height as the bay widened, but it still overtopped the island, which rose to 150 feet at its highest point. Ulrich couldn't get the anchor completely up, but he managed to turn the boat around so that it was facing the oncoming water, which would give him a better chance to ride it out.

When it did reach him—perhaps three minutes after the initial splash, Ulrich thought—the wave was about seventy-five feet high. He gunned the engine and headed into it. His anchor chain snapped as the boat rose up. At one point he looked down on the shore and thought that if the wave broke he'd end up in the treetops. But it didn't break, and instead he rode up and over the crest and down the backside, into the middle of the bay. The wave kept going, across the sand spit that the mountain climbers had vacated about an hour before.

The two other boats in the bay were not so lucky. The wave swamped them both. The crew of the *Badger* were able to get into their skiff and were picked up later; the crew of the *Sunmore* were lost. Ulrich's boat was tossed about for twenty minutes as the water rebounded from shore to shore, but ultimately the bay calmed down.

The next day, Don Miller—the geologist who taught George Plafker so much about working in backcountry Alaska—flew over the bay. He'd studied wave-related damage there some years before, occasionally with Plafker, so when he heard about the earthquake he arranged a quick overflight. Intrigued by what he saw, he visited on the ground a month later, climbing up the hillside opposite the rockfall. The splash had acted like a giant scrub brush, scouring the vegetation off the hillside. This was not unusual; Miller and others had seen this phenomenon before, here in Lituya Bay. There was a term for the highest extent of the scouring: the trimline.

What was astonishing this time was the trimline's height. Miller had lugged an altimeter with him, and when he reached the highest point of the trimline, the instrument read 1,720 feet above sea level. The splash that occurred that night in July was, and still is, the highest wave ever documented anywhere on earth.

Of the thousands upon thousands of earthquakes that happen around the world every year—from imperceptible tremors to powerful shakers like the one that hit Lituya Bay—roughly one in sixteen occurs in Alaska. That makes the state one of the most quake-intensive places on the planet. For a long time, however, no one knew that. In part that's because for most of history, Alaska was nearly devoid of people. But it's also because for a long time no one knew much about earthquakes.

In ancient times, the shaking of the earth was often attributed to the movements of a giant mythical beast. In Japan (another earthquake-intensive part of the planet) that beast was a catfish, thrashing about. In some Native American storytelling, it was a tortoise that carried the world; when it took a step, the earth shook. Once societies moved beyond myths, there were attempts at more-scientific explanations. Earthquakes often accompanied volcanic eruptions, and eruptions usually involved the release of large amounts of gas. So there was some logic to the idea, put forth by Aristotle in the fourth century BC, among others, that earthquakes were somehow related to the movement of gas underground. The concept took many forms over many centuries. Some thinkers believed that earthquakes were caused, not by gases, but by winds in caverns deep within the earth. Others thought they might be related to the *collapse* of caverns after volcanic gases within them had escaped. In the seventeenth century, the French philosopher-scientist René Descartes wrote that rather than simply moving, perhaps gases actually exploded underground to cause earthquakes. Even a century later, winds or steam or other gases were still the favorite earthquake-causing mechanism of many scientists.

The idea began to fade in the nineteenth century, for a couple of reasons. Largely as a result of several major earthquakes—one that devastated the Portuguese capital of Lisbon in 1755 and a series of five that struck the region of Calabria in southern Italy in 1783—scientists had begun to study seismicity more closely. As they learned more about where quakes occurred, they realized that some happened in places where there was no volcanic activity. And with many quakes they began to notice evidence that the earth had broken—that one piece of the ground had moved in relation to another. The place where this movement occurred was called a fault. Sometimes the movement was horizontal, creating what appeared to be a line in the earth as one block of earth slipped past an adjoining one, and sometimes it was vertical, creating a scarp, or cliff, as one block of earth rose above the other.

In places where faulting was visible, the question then became a chicken-and-egg one: was the fault simply the result of the shaking that happens during an earthquake, or was it the source of the shaking, the origin of the earthquake itself?

For a long time in the nineteenth century the answer was that an earthquake occurred somewhere else and that the movement of the earth on the fault was only a consequence of the quake. Contraction theory—the idea that the earth was cooling, and shrinking as it did—proved useful here, as it did in rebutting continental drift. Earthquakes could occur when parts of the crust sank during contraction, the thinking went, leaving breaks in the surface as an indication of the movement.

As the century progressed, though, scientists began to see the earthquake–fault connection in the opposite way. Of nineteenth-century geologists who came to this conclusion, no one stated it more clearly, and perhaps with more authority, than Grove K. Gilbert.

Gilbert was born in upstate New York in 1843 and home-schooled until he went off to college at the University of Rochester. After graduating he got work with a company that supplied specimens to museums. It was while excavating a mastodon at a waterfall outside of Albany that Gilbert became interested in landforms and how they changed over time. It was an interest that directed and informed his entire career.

Itching to do geological fieldwork, he soon landed a job with the Ohio Geological Survey, and two years' work there eventually got him a recommendation to join one of the great western surveys, being led by an army lieutenant, George Wheeler. As Wheeler's geologist in the early 1870s, Gilbert explored New Mexico, Arizona and Utah, then joined another of the great survey leaders, Major John Wesley Powell—who had led the first expedition down the Colorado River—for several more years. He later followed Powell to the fledgling US Geological Survey as its first chief geologist.

On the basis of his explorations with Powell in the Rockies, Gilbert wrote a report on the Henry Mountains in southeastern Utah, considered a masterpiece of thinking and writing about geology, especially the role that water plays in altering the landscape. It was in a lesser-known report, however, that Gilbert set forth his ideas about faults and earthquakes. In his explorations in the West and his thinking about how mountains came to be, he had come across clear signs of faulting in the form of scarps in the Wasatch Mountains of Utah. These spectacular sheer cliffs were created during the process of mountain building, when blocks of crust moved upward. In 1884, in a short report that warned Salt Lake City of the danger of potential earthquakes, Gilbert wrote that "upthrust," as he called it, distorted and compressed the crust, building up stresses until the force was enough to overcome the friction that keeps one block fixed to another. Then one of the blocks shot upward. "Suddenly, and almost instantaneously, there is an amount of motion sufficient to relieve the strain," he wrote. "This is followed by a long period of quiet, during which the strain is gradually reimposed."

Gilbert had succinctly stated the basic mechanism of earthquakes. It took another major quake, and the investigative work that followed it, for the mechanism to be described in detail.

That quake was the great San Francisco earthquake of 1906, which occurred along the San Andreas Fault. Unlike the vertical scarps that result from mountain building, the San Andreas moves horizontally. But the same rule applies—stresses act on the fault until they become so strong that it breaks and one side suddenly slips past the other. The fault eventually quiets down and the stresses start to build up again.

The San Andreas Fault had been recognized for a long time—it is visible on the surface along much of its 810-mile run from Cape Mendocino in Northern California to the Salton Sea in the south. And for four decades beginning in the 1850s, measurements had been made on either side of it, showing that it was gradually deforming—the western side of it was slowly moving north relative to the eastern side. After the 1906 quake, which was centered just off the coast south of San Francisco, new measurements showed that part of the fault had suddenly moved a great distance—in some places twenty feet or more.

Harry F. Reid, a geophysicist at Johns Hopkins University and a member of the scientific commission that investigated the earthquake, formulated a theory to connect the ground movement to earthquakes. Called the elastic rebound theory, it proposed that forces in the earth caused the rocks on either side of a fault to absorb energy and deform—just as the two sides of the San Andreas Fault had slowly deformed over the years. Once the stresses exceeded the rocks' internal strength, the fault broke suddenly, releasing much of the stored energy. It was this release of energy that caused the ground shaking and destruction.

Reid's theory was gradually accepted over the next several decades, as it helped scientists answer a lot of basic questions about earthquakes. For one thing, it accounted for the varying strength of different quakes: the longer the fault that broke, the stronger the quake, because more energy had been stored and released.

But the theory did little to answer perhaps the most fundamental question about earthquakes. What was the source of the forces that were acting on the faults, deforming them until they broke? Where was all that energy coming from? The answer to that question would come later.

Along with a greater understanding of the mechanism of earthquakes, by the late nineteenth century scientists were gaining more knowledge of how often and where they occurred, using increasingly sophisticated seismometers, instruments that could detect the shaking of the earth, often from far away.

By then, seismometers had been around in one form or another for more than 1,700 years. The earliest-known one was a Chinese invention from the second century. It consisted of a large bronze jar with dragon heads arrayed around it like the points of a compass, a bronze toad beneath each dragon and most likely some kind of pendulum inside. Each dragon held a ball in its mouth. Shaking from a certain direction would cause the ball from one dragon to drop into the mouth of the accompanying toad.

In Europe, seismometers were first developed in the eighteenth century, with great improvements coming after the Calabrian earthquakes of 1783. In addition to direction, these instruments could give some idea of the power of a quake. But it wasn't until near the end of the nineteenth century that scientists developed reliable instruments that could measure ground motion over time—what are now called seismographs. The pioneering work on these instruments was done by British scientists working and teaching in Japan.

With a seismograph providing a time frame, it was now possible to determine, more or less accurately, the point of origin of an earthquake—its epicenter—by comparing the arrival times and directions of seismic waves. Scientists now located earthquakes in all sorts of places—in isolated mountain ranges, or deep under the sea. Throughout the first half of the twentieth century, as instruments became even more sensitive and more broadly distributed around the world, and as communications and data processing improved, seismologists were able to compile lists of thousands upon thousands of earthquakes, including data about their strengths and locations.

By the 1940s, several things became apparent. First, scientists noticed that the frequency of earthquakes is related to their size—that is, powerful earthquakes are rare, and weak earthquakes happen all the time. We now know that on average there is about one "mega" quake per year, while there are perhaps one hundred thousand quakes that are just strong enough to be felt by someone not too far away, and more than a million that are below the limits of perception of most people.

But of more immediate interest to the scientists who were becoming convinced that the continents moved—that Alfred Wegener was right, though he may have had the details wrong—was what seismologists learned about the geographic distribution of earthquakes. When quake epicenters were plotted on a map of the earth, several earthquake belts became apparent. Some of them were in the middle of the oceans—one ran north to south in the middle of the Atlantic, and another through the eastern Pacific. This correlated perfectly with the midocean ridges that scientists at Lamont Geological Laboratory had mapped using depth-finder data, beginning in the late 1940s.

But the largest number of earthquakes ran around the rim of the Pacific Ocean, from Alaska southward along the western coasts of North, Central and South America and then northward through Oceania, Asia and Siberia. This was the same belt that was teeming with mountains and volcanoes. It was now clear that whatever was making those geographic features was also responsible for the earthquakes. Scientists still hadn't yet figured out what "whatever" was, but by the 1950s they were getting tantalizingly close.

As seismologists learned when they started compiling more comprehensive lists of earthquakes around the world, Alaska has more than its fair share. Not only does the state have a lot of earthquakes, it has a lot of strong ones. One study in the 1940s suggested about 7 percent of all the earthquake energy released in the world each year was released in Alaska. The quake in Lituya Bay on that July night in 1958 was one of the bigger ones Alaska had experienced, but it was hardly unique.

Records of quakes prior to the twentieth century are sketchy, given the lack of instrumentation and the nearly complete lack of inhabitants. The first two reported earthquakes in Alaska, noted in accounts by Russian traders, occurred in July and August 1788 and created tidal waves that struck Kodiak Island and smaller islands in the Gulf of Alaska. The August quake was especially destructive, killing an unknown number of natives and drowning hogs on Sanak Island, off the Alaska Peninsula.

Two more large quakes were reported in the mid-nineteenth century. One, in 1843 near Sitka, deflected the needle on a device used to measure magnetic field intensity. It is thought to be the first Alaska quake to be detected by an instrument, even if the instrument wasn't a seismometer. Sitka was struck again in 1847; although there is no record of damage or loss of life, this quake was thought to be the largest of the Russian period.

Earthquake reporting picked up after the United States' purchase of Alaska in 1867, as volunteers with the Weather Bureau of the federal Department of Agriculture were encouraged to report any shaking going on. Many small earthquakes were reported after that time. Then, on September 10, 1899, a series of very large quakes—including two within thirty-seven minutes of each other—struck Yakutat Bay, on the southeastern edge of the Copper River delta. The shaking was felt from Fairbanks, in the north, to Sitka, down the coast, a distance of six hundred miles, and was strong enough near the epicenter to throw people off their feet, although no one was killed. The shocks also shattered Muir Glacier on Glacier Bay, about one hundred miles to the southeast. The front of the glacier was a popular destination for tourist boats, but its destruction discharged so many icebergs that the bay essentially became unnavigable for a decade.

The 1958 earthquake was the biggest since 1899. In addition to the two people who died in Lituya Bay, three people were killed on Khantaak Island at Yakutat Bay. Jeanice Welsh, who may have been the first woman to own a cannery in Alaska, had gone to the island with two friends to picnic and pick wild strawberries. The three disappeared when the land slumped into the water.

The five deaths serve to illustrate that while people fear the severe ground movements that can accompany an earthquake, sometimes the biggest, most spectacular and most deadly effects are only indirectly related to the shaking. But terrible as those effects were in 1958, they were only a modest foreshadowing of events that would occur six years later, when, at 5:36 p.m. on March 27, 1964, Alaskans experienced a quake unlike any that had come before.

9

SHAKEN

The motion was gentle at first: a slight rumbling and rolling, like being on a slow train ambling down a rickety track.

It was felt across a wide swath of Alaska, a great half circle of nearly half a million square miles bordered by the Gulf of Alaska to the south, and arcing from Fort Randall, at the western tip of the Alaska Peninsula, through Bethel, on the Kuskokwim River, in and around Fairbanks, in the interior, across the Canadian border to Dawson, in the Yukon, and down to Skagway, in Alaska's southeastern Panhandle.

Alaskans were accustomed to earthquakes, and when the shaking started the almost universal thought was that it would soon be over. In Anchorage, where many youngsters were glued to the television before dinner, watching an episode of the sci-fi marionette series *Fireball XL5,* parents told them not to worry. The danger would pass, just as it had many times before. At her medical office south of downtown, Dr. Louise Ordman also thought it was just another small tremor. So did Dean Smith, sitting in the operator's seat of a gantry crane fifty feet above the docks in Seward, on Resurrection Bay in the Kenai Peninsula.

For some, the first thought was that it must be something other than a quake. Tobias Shugak, a young boy in the native village of Old Harbor on Kodiak Island—where Kris Madsen had taught the previous year—thought his sister was shaking the bed he was lying on. It seemed to Irving Wedmore, out fishing in Unakwik Inlet in Prince William Sound, that his boat had run aground. Other boaters thought their propellers had become entangled in something. At the city dock in Kodiak, one fisherman was sure his craft had been hit by another one.

After about thirty seconds, though, the movements grew much stronger and much faster. It was right about then that Alaskans began to see and hear things they'd never seen or heard before.

The rumbling became a violent jarring shake, and the rolling increased in intensity so that the land began to resemble the sea, as shock waves rippled through the ground, pavement, buildings. Anyone standing started to have trouble remaining upright, and many fell or were thrown to the floor by the sharp motion. Walking became nearly impossible; for those who managed to make some headway, the effect was like that of a circus clown slipping and sliding on a barrel. Ordman, staggering out of her building through the lobby, was heaved violently against a door frame, first on one side and then the other. The top of Smith's crane started swaying back and forth like a whip; as the entire machine hopped around on the dock ("like a stiff-legged spider," he recalled later), he scampered down to safety. In Old Harbor, Tobias Shugak watched as the small houses seemed to dance in place and as the pilings of the dock rippled in the water.

Many heard a deep roar as the quake got up to speed. A woman in Anchorage who was heating a pot of tomato sauce for that night's dinner thought a valve on her stove had broken, because it sounded as if the burner had jumped from simmer to full blast on its own. A young man on a ski slope next to the runway at Elmendorf Air Force Base was certain a fighter jet was coming in for a landing; when he looked up and didn't see one, his next thought was that the army base next door was having artillery practice.

For others, the deep sounds were overwhelmed by the sharper and closer sounds of glass breaking, nails popping and wood splintering. Away from the cities and towns, some heard sharp cracks as thick ice shattered on lakes, or loud booms followed by a distant rushing sound as snow and rocks broke and tumbled down mountainsides. This being the height of the Cold War, many Alaskans who heard a booming sound thought that the Soviet Union must have dropped an atomic bomb on their state. In his cabin outside of Cordova, one woodsman was so convinced Russian battleships were shelling the coast that he grabbed his hunting rifle, hopped into his truck and drove toward town to fight the expected invaders in the streets.

On the wide boulevards of Anchorage, drivers thought there was something wrong with their cars. Many believed that the problem had to be a flat tire—the car was lurching and seemed out of balance. Some figured an entire wheel had come off; to others, it seemed as if the rear axle had separated, taking the whole back end of the car with it. As vehicles jumped and slid—side to side or forward to back, depending on their orientation—many drivers tried to steer their way straight. Something had happened to the steering wheel, they thought. Others jammed on the brakes, to no avail. An Anchorage taxi driver, Joe Kramer, was convinced that *his* car wasn't the problem. All the drivers around him were acting crazy, he thought.

Inside homes and apartments, kitchen cupboards opened and closed, the contents spilling out. Latched refrigerator doors broke open. Cabinets toppled over, the china within reduced to shards in seconds. Television consoles fell and books tumbled from shelves. Doors became jammed as their frames twisted and torqued. Stairways wriggled and writhed, and furniture that managed to remain upright moved across rooms. In one Anchorage house, a six-drawer dresser left scratch marks as it meandered like a drunk across the asphalt-tile floor. The markings later proved invaluable to scientists studying the ground motions of the quake.

Those who had a vague idea about how to protect themselves got under archways or in doorways, or tried to crawl under furniture. One college student somehow managed to slide under the bed in her dorm room, although it was only seven inches off the floor. The shaking "scared the curl right out of my hair," she said.

Inside or outside, objects swayed crazily from side to side. At one Anchorage home, a chandelier swung so violently it hit the ceiling beams. As the earth rolled beneath them, trees slapped back and forth like the windshield wipers on a car. Top-heavy trees couldn't take the strain and snapped off high on their trunks. Telephone poles and streetlights whipped back and forth, the wires connecting them alternately going taut and slack or breaking under the tension.

The mayor of Anchorage, George Sharrock, was driving near the airport when he saw a raven trying to land on top of a light pole. But the pole was not cooperating, swaying out of the way every time the bird tried to put its claws down. Eventually the raven gave up and flew off.

At the airport itself, Chris von Imhof, the local manager for Scandinavian Airlines System, tried to leave his second-floor office in the terminal, but the nearest exit doors were locked. He kicked on a window to break it, lowered himself as far as he could and jumped the rest of the way. As he hit the ground he looked up and saw a terrifying sight: the six-story control tower next to the terminal cracked and then crumbled onto itself. Von Imhof spent the next hour digging through the rubble with others, and managed to pull two cooks, alive, from the Northwest Airlines kitchen that had been on the first floor. Shortly after that they uncovered the body of William Taylor, the air traffic controller who had been on duty at the top of the tower. At one point while they were frantically working, a plane descended through the overcast skies toward the runway, which had heaved so much that it now had large cracks running through it. Like the raven, the plane gave up and flew off.

Closer to downtown, Glen Faulkner, a geologist, ran out of his house and looked down the street where the Four Seasons, a six-story apartment building, was under construction. Just an hour earlier it had been crawling with workers. The structure was being built using a technique called "lift-slab"—the concrete floors had been poured on the ground on top of one another, like a stack of pancakes, and then jacked into position one at a time. Faulkner saw that the building was shaking severely, in a north–south direction. After a minute or two it tilted slightly, slid a bit to the north and collapsed, the floors falling one onto another as they had been raised, one by one. A concrete elevator shaft was the only thing left standing, and it leaned precariously at an extreme angle.

The ground was still moving. A driver in Anchorage saw a hundred-foot-wide variety store rolling up and down as the seismic waves passed through it. It looked like a caterpillar, she thought, and amazingly, it remained intact. Another driver rode her car as it bounced up and down on a street that was, as she described it, "writhing like a snake."

The seismic stresses caused cracks to open in the ground all over southern Alaska. Some of these grew larger and larger as the shaking continued. They opened and closed as the waves rolled through, forcing liquids into the air as they did. The effect in some places was like a choreographed fountain—curtains of mud up to fifty feet high and one hundred feet long, spewing for a few seconds as the trough of a wave passed and the ground closed up, then stopping as the crest followed and the ground opened up again.

In Portage, along Turnagain Arm, the postmaster reported afterward that the average crack was a quarter mile long and six feet wide. But there were many smaller cracks as well, and as the ground in some places split into a crazy-quilt pattern other bizarre effects were noted. Two men at a service station in Portage ran out of the building as the quake began, and once they were outside, a three-foot crack opened between them. Then smaller cracks formed around each of the men, so that in a few moments they were both standing on isolated islands of earth. As the seismic waves kept coming, the two islands began moving up and down. One of the men described it as like being on an open-air elevator: as his block of land went up, his friend's went down.

On frozen Portage Lake, a long and deep body of water that was formed and fed by the nearby Portage Glacier, Ruth Schmidt, a geologist and lecturer at Anchorage Community College, had been making depth measurements with four other researchers. She was on skis about four hundred feet from the northeast shore of the lake when there was a jolt and the ice—two and a half feet of it, topped by more than three feet of snow—began to move. She started skiing toward her colleagues in the center of the lake, and felt the ice undulating beneath her the whole time. Her ears were filled with loud roaring and creaking sounds. As the quake wound down and the five researchers headed together toward shore, they saw a six-foot-high ridge of broken ice near what had previously been a flat shoreline. The whole ice sheet on the lake—two miles long by nearly a mile wide—had moved back and forth, colliding with the land, and the fractured ice had had nowhere to go but up.

In downtown Anchorage, Blanche Clark had just left the J. C. Penney department store on Fifth Avenue. She worked part-time as a courier for a photo processor, and Penney's was one of her usual late-afternoon stops. The store, opened just a year before, was a retailing marvel, full of clothing and just about anything else that a modern Anchorage family might need on five floors connected by escalators. From outside, it was an imposing structure, windowless save for the street level. Like many modern buildings, it had a facade made from decorative prefabricated concrete panels attached to the building's steel frame.

Clark came out of the store with a bag of film for processing and got into her 1963 Chevy station wagon, which was parked on the street along the north side of the building. She started the engine and was waiting for the traffic to clear before pulling out into the street when the ground began to shake. She thought about getting out of the car but saw that a woman crossing the street was having trouble walking. Clark decided she'd ride out the quake in her car.

Inside the store, Carol Tucker was on the third floor, browsing the bedding and china departments. The lights quickly went out, leaving the floor in almost complete darkness. Things were falling from the ceiling around her, so Tucker crawled to where she thought the escalators were, and found one. Barely able to stand up, keeping her hands over her head to protect herself, she started down the metal steps—and promptly fell most of the way to the second floor, tearing ligaments in her leg. The second floor was heaving up and down, so despite the pain, Tucker kept going down the next escalator. When she reached the first floor, she made her way toward daylight and the exit. But rather than hobbling outside, she hesitated. Looking back later she could not explain why she paused, but by doing so she probably saved her life.

As Tucker stood there and the severe shaking continued, the building's exterior panels started to break off and fall. One landed on the sidewalk just outside the door where Tucker stood. Another struck the woman on the street whom Blanche Clark had been watching just moments before, cutting her in half. Others fell onto empty cars parked up and down the street. Then one landed on Clark's station wagon, flattening its roof under several thousand pounds of concrete. Clark was wedged into the seat, her shoulder and chest screaming in pain. But she was alive.

One block over, on Fourth Avenue, a different kind of hell was breaking loose. The street, at the crest of a hill that slopes northward down to Ship Creek, started to slide as the shaking continued. That caused Fourth Avenue to crack down the middle for several blocks. The north side of the street then dropped about ten feet, taking businesses with it, including the Anchorage Arcade, Mac's Foto, Pioneer Loans and three bars. At one of them, the D&D Café, several regulars had been playing cards when the shaking began. They rode the building down and then, when the sinking and shaking stopped, put their cards down and crawled up and out to safety.

That was one of several slides in Anchorage. But it was far from the worst.

In the living room of his log home in Turnagain-by-the-Sea, a subdivision built on a bluff overlooking Cook Inlet near the airport, Bob Atwood was practicing the trumpet when the quake began. The owner and publisher of the *Anchorage Daily Times,* the state's largest newspaper, Atwood, fifty-seven, was one of Alaska's power brokers. He'd used the pulpit of the press to push for statehood (and had been appointed chairman of the Alaska Statehood Committee when it was formed in 1949), for more military spending, for the expansion of Anchorage's airport and for any number of other state and city causes. He and his wife, Evangeline, were major contributors to the arts in Anchorage as well.

Atwood had just gotten home from the newspaper's downtown offices, and Evangeline was off shopping for groceries in Spenard, across Chester Creek from Anchorage proper, for the dinner party the couple had planned for that evening. It was a perfect time to practice, since he could blow the trumpet loudly without disturbing anyone.

No sooner had he put the instrument to his lips, though, than the house started rocking. A chandelier, hanging from a beam in the living room, began to sway. Soon the whole house was lurching about, as if it were being heaved to and fro.

It was obvious to Atwood that the house wouldn't stay in one piece for long; the large roof, in particular, seemed in danger of caving in, given how every other part of the structure was bending at odd angles. He ran out the door and down the driveway.

When he stopped and looked back, the ground under the house was moving, stretching the structure apart at one moment and compressing it the next, as if it were a giant squeezebox. But that didn't last long, as the forces of the roiling earth soon became too much for the house. It broke apart to a terrible noise of glass cracking, huge logs splitting and the house's contents being crushed and crumpled.

Getting out when he had had saved his life. But Atwood didn't have much time to think about that, or about the loss of his worldly possessions. Around him trees were falling over. Worse, the ground itself was starting to break into strange, angular blocks, some rotating up and others down. It was as if swarms of organisms were inside the soil, giving it life. Atwood began to wonder if he would be able to stand anywhere.

Suddenly a crevasse opened beneath his feet, and he was falling. To Atwood it seemed that he fell a long distance, and although it was still light out suddenly he was in darkness. But he landed in sand, miraculously soft and dry. He saw that he was in a deep V-shaped chasm, and it was starting to fill up with other objects—tree stumps, fence posts and boulder-sized chunks of frozen soil. His right arm seemed to be buried in the sand, and he realized that his right hand was still holding his trumpet. Whole trees were now falling into the crevasse, which was getting wider and growing laterally toward the house of Atwood's neighbor, Lloyd Hines, an optometrist, and his family. Atwood could see their house through the ever-lengthening chasm. It appeared to be sliding toward him.

After a while—he wasn't sure how long—the house stopped moving. In fact, Atwood realized that everything had stopped moving, or at least was only moving slightly. It was quiet, except for the occasional crash of damaged structures or trees. He let go of the trumpet, wriggled his arm free and slowly climbed out of the crevasse.

The neighborhood was gone. Some houses were reduced to splinters, heaps of broken wood and glass. Others were still intact but were scattered this way and that, like dice rolled on an uneven surface. A few were nearly upside down. Atwood's house, which had been about seventy feet above the water, was now down at sea level, a pile of kindling except the roof, which amazingly was in one piece. It would be used as a landing pad for rescue helicopters the next day.

The birches and cottonwoods that had characterized the neighborhood were largely flattened. Atwood, still in his suit and tie from the office, looked around for others who might have survived. He heard two of the Hines children, eleven-year-old Warren and four-year-old Mitzi, crying. Climbing over the jumbled terrain, he eventually reached them. Then another neighbor—a mother with her four children—appeared. Together, the eight of them set off to find stable ground.

Soon they came across Margaret Hines, Lloyd's wife. She had been driving in her car and had just turned into the driveway when the quake started. The land around her had shaken and broken up and fallen away, taking the house with it, and she and the car had been left perched on an isolated patch of ground, a giant earthen toadstool high in the air. Seeing the other survivors, she climbed down and joined them.

Atwood looked around to try to figure out which way led to solid ground. In one direction he saw a stand of trees that appeared to still be vertical—a good sign, he thought. The motley group headed that way, with Atwood in the lead, scouting for a safe route. It was slow going, climbing up and down over blocks of earth and broken trees, but eventually they could see a bluff, with people standing on it. It was a new bluff, apparently, created when all the land to seaward—the land that his house and countless others had sat on—slid away. As Atwood and his companions neared it, a rescue party came down, with ropes and saws, to help them climb to safety.

When the shaking started, Kris Madsen knew instantly what it was. She was from Southern California, after all, and she recognized an earthquake when she felt one.

She was up at the pond behind Chenega's schoolhouse, with Norman Selanoff, fetching water. The ground started to move, and like so many others that day she thought, Okay, I've been through these kinds of things before, it'll stop.

But the shaking soon got so bad that Madsen could barely stand up. Then she realized that the trees around her—mostly large spruces—were swaying. But it wasn't just the tops that were moving, as would be the case in a windstorm. Entire trunks were swinging from side to side, from the ground up, like a metronome.

In the midst of this she turned around and looked out. The view from the hill was still magnificent, the mountainous islands stretching out in the distance beyond Whale Bay and Bainbridge Passage. But closer in, something was startlingly wrong. The water in the cove had disappeared. It was as if a giant vacuum hose had come along and sucked it dry. She had never experienced anything like this.

Down in the village, nineteen-year-old Avis Kompkoff had taken little Lloyd, her infant, out of his baby seat and put him on the bed. She had gotten him dressed and put booties on his feet. Then she thought about her bath. She'd need fresh clothes afterward, so she leaned over to get some out of the bottom drawer of her dresser.

It was then that her three-year-old, Joey, jumped up on the bed. He didn't speak, but he seemed spooked: there was a wildness in his eyes. What the hell is the matter with him? Avis asked herself.

Then she felt the shaking.

She wasn't unfamiliar with earthquakes, either. Just a few weeks before, in fact, there had been a mild quake, a rumbling that had lasted a few seconds. But this one was different, Avis knew: it went on and on, and the motion became so severe that the little house seemed as if it might fall apart. She didn't know what to do but remembered something that older people in the village had once told her: in an earthquake, get the door open if you can. Otherwise, if the shaking is strong, the house can quickly go out of kilter, the door will become jammed and you'll never get out. Avis decided to go her elders one better: open the door *and* get out. She grabbed the two boys and went outside.

She found herself next to a neighbor, Steve Eleshansky, whose house was just uphill from hers. He had gone outside, too, holding his one-year old daughter, Rhonda. His wife, Dorothy, known to everyone as Tiny, was nowhere to be seen, as was their other child, five-year-old Steve Jr.

In a few moments Avis's husband, Joe, who had been down below the bulkhead on the beach, came running up. Like Madsen, Joe had seen the cove become suddenly, and eerily, empty. Even more ominously, a moment or two after that, he'd heard someone shouting about a tidal wave. He took Joey and told Avis to bring the baby and follow him.

The Kompkoffs' house was on the south side of the village. To reach the steps to the schoolhouse they would have had to run along the top of the bulkhead to the store. They didn't have time for that. There was a hill directly behind their house, separated by a ravine from the hill the schoolhouse was on, with a narrow rope bridge connecting the two. They went straight up this closer hill, with Joe carrying Joey and Avis carrying Lloyd. Behind them, lagging somewhat, was Steve Eleshansky, carrying Rhonda. Avis quickly lost her slippers as they struggled up the slope through the waist-high snow. The terrain was open at first, but in a short while they reached a stand of trees. Avis briefly got one of her feet trapped in some tree roots, freed it and kept going.

Nick Kompkoff's first thought when the earthquake started was about his oil-burning stove at home. His wife, Mary, had been cooking a pot of chili on it that afternoon. The stovepipe was unstable even in the best of circumstances, and he was worried that the whole thing would topple over and set the house on fire. Kompkoff had been on his way to the Smokehouse to shoot some pool, but now he hurried down the boardwalk toward his house on the opposite side of the village. Before he got there he ran into Mary, who was wondering about the children. Nick Jr. and the other boys had come back from the beach a while before; they'd gotten thirsty, so they'd stopped in at their Aunt Shirley's house for water. But the three girls—Carol Ann, Julia and Norma Jean—were still down at the beach. Nick looked in that direction and saw them on the dock.

He started running toward the girls, although by now the boardwalk was moving so violently he could barely stay on it. As he reached the dock he noticed the water receding from the cove.

He reached the girls and picked up the younger ones, Carol Ann and Norma Jean, one in each arm. He shouted at the oldest, Julia, to come along. They ran toward land and the safety of high ground.

At the far end of the beach, Timmy Selanoff watched in amazement as the rocks that he had been walking past began to bounce, like the ball in a game of jacks. He heard his friends calling to him from far away. But he couldn't make out what they were saying, the earth was rumbling so loudly.

He knew he had to get back to the village, so he started jumping from boulder to boulder along the shore, nimble despite all the ground movement. The stones in his jacket pockets, his "ammunition" against the birds, were weighing him down, but he didn't think to empty them. He was scrambling for his life.

Recollections of survivors in the days that followed suggest that the first wave of water arrived in the cove less than a minute or so into the earthquake. It was more like a fast-rising tide than a wave, though, and since the tide was low to begin with, it came only about halfway up the beach. The incoming water was quiet, or it seemed so amid the shaking and rumbling.

That first wave rapidly retreated, and then some. This was the vacuum effect that Madsen and the others saw. The bottom of the cove was exposed out to a distance of about a quarter of a mile and a depth of more than 120 feet. It was as if a canyon had instantly been revealed.

Then, about two minutes later, a bigger wave came in. Survivors estimated it was a wall of water thirty-five feet high when it approached the shore.

The ground was still shaking at that point. Up on the hill, Madsen was struggling to walk in the snow. She and her friend had decided to head farther uphill, away from the schoolhouse. They were not sure how much longer it would be standing.

Just then Madsen heard a tremendous noise from below. She turned in time to see the second wave crashing into the village. The water seemed to drop onto the houses and the church, the Smokehouse, the bathhouses—everything. She could hear the sounds of homes being shoved off their pilings and broken apart, of the church splintering, of the dock breaking, of boats being lifted and tossed about, of trees being snapped. And amid all the unnatural noises were the unmistakable sounds of villagers shouting and screaming.

Kenny Selanoff, who had been playing down on the beach, managed to make it up to the village before the big wave struck. The thirteen-year-old dragged his younger brother George with him as he scrambled up the slippery wooden steps of the bulkhead. Once in the village, they grabbed on to a clothesline pole and, amazingly, managed to hold on while the water surged around them. Kenny saw homes being destroyed—among them the house of Avis's adoptive parents, Willie and Sally Evanoff, and her daughter Jo Ann. According to Kenny, the three were standing in their doorway when the wave hit. It pushed them inside, and they were gone.

Avis's aunt, Margaret Borodkin, had been trapped inside another house when the structure partially collapsed during the shaking. Her hip and leg were badly injured, and she had been pinned to the floor. Then the second wave had struck, shattering the house into pieces as if it had been hit by a bomb, she recalled later. She passed out as the waters churned around her.

That second wave surged across the flat expanse of the village, pushing some of the debris into the hillside. With nowhere to go but up, the water rose rapidly toward the schoolhouse, swirling around in a violent foam of debris, mud and silt. It wiped out the ninety steps to the top, as well as the rail system next to it that had been used to deliver supplies to the school. The water kept rising, reaching the play yard just below the schoolhouse and inundating the newer generator shed and the empty wooden workshop building next door that had housed the generator in the past. It rose some more, until it reached the foundation of the schoolhouse itself, seventy feet up. Then it stopped.

After a few moments, the water started to subside, and then retreated back into the cove—taking the village, now reduced to a sad jumble of broken homes and other debris, with it. All that was left of most of the homes were the pilings they had stood on. Where the dock had been, the pilings looked as if they had been leveled with a chain saw.

Joe and Avis Kompkoff and their two children had beaten the wave, making it up the hill behind their house and reaching the schoolhouse by scurrying across the footbridge, which was swaying crazily. With the water at its highest point, at last Avis turned around. Steve Eleshansky and his daughter Rhonda were nowhere to be seen.

Nick Kompkoff heard the shouts of "Tidal wave!" too. He glanced back and started running faster, urging Julia on as well. But they couldn't outrun the water, and it was soon upon them. As he described it several years later, he reached out in a vain attempt to grab hold of Julia. In doing so, Norma Jean came out of his arm. The water swept both of the girls away. Julia called out to him, screaming "Dad!" It was the last word he ever heard from her.

He was in the water, too, desperately holding on to Carol Ann. The girl was wearing a hooded parka that was zipped closed. Kompkoff never figured out exactly what happened next, but he ended up at the back of the village, in a snowbank. He heard a loud cracking noise, and something—it might have been a large pole—hit his back. He passed out, and when he awoke he was holding on to a log or other piece of debris with his right arm, his back was aching and he had something in his left hand. He looked at it: it was the hood of Carol Ann's parka. He tugged and felt the weight of his daughter, who was still zipped inside it, and still very much alive.

For Timmy Selanoff, the village was too far away, and the water was coming too fast. He had to get away from it, and the only way was up—nearly straight up, since where he was on the beach a steep hill came almost straight down to the water.

Timmy started up. He remembered grabbing on to a twig or root at some point. Other than that, he recalled later, he had no memory of what happened. He credited divine intervention—as the second wave approached, a voice had told him not to be afraid.

When the shaking finally stopped, the schoolhouse was still standing, a testament to the strength of its concrete foundation. Madsen and Selanoff went down the hill to it and went inside. Water had entered the basement, but had gotten no farther. There was plenty of other damage, though. Jars of surplus food had fallen off the shelves in the storeroom and were now broken on the floor. Butter, peanut butter and other food was everywhere, mingled with broken glass and books.

Avis and Joe Kompkoff and their two children had been among

the first of the villagers to make it to safety. Over the next few hours, by ones and twos, more survivors straggled up the hill. Nick Kompkoff, with his badly injured back, managed to crawl up the snowy hillside. About halfway up he met some villagers, who took Carol Ann and brought her the rest of the way. But Kompkoff heard a woman crying from below, so he rolled back down to where the voice was coming from. It was Tiny Eleshansky, in the snow. The raging water had ripped off her clothes save one sock. Ken Vlasoff and John Brizgaloff had heard her cries too and had come down the hill to help. Together, the three men brought her up to the school.

Almost everyone, it turned out, needed something—they had fled the water with only the clothes on their backs and had lost everything. Madsen, overwhelmed and numbed by what she had seen, did the only thing she could think to do—begin handing out whatever was available in her apartment. She got a blanket and some clothes for Tiny Eleshansky. She gave Avis Kompkoff a pair of shoes to replace her lost slippers. Someone else got her rabbitskin coat, the one Madsen had bought two years before with her mother in Anchorage. Soon she had given out all of her spare clothing and shoes.

As the villagers gathered, in shock, some moaned in pain and others cried out to no one in particular for help. Some tried to take stock of who was still missing. Parents called out for lost children, and children screamed and cried for their parents. There were tearful reunions. But there were far more tears for reunions that didn't happen.

The occasional rattling aftershock spooked the villagers enough that they decided to get away from the school, in case it were to collapse. They moved uphill toward the pond and set up a temporary camp of sorts amid the spruces. Someone made a bonfire, and people used cardboard boxes from the school, folded flat, to insulate themselves from the snow. Despite the damage inside the schoolhouse, there was plenty of food for those who were hungry, and milk for the children.

The search for the missing continued, with some of the men going back down the hill to the flat expanse where the homes had been, calling out names. One house, out of the direct path of the water on the far eastern side of the village, had been spared, but everything else was gone.

While they were searching through the devastation, they heard a voice out in the water. It was a woman, obviously suffering from pain and fatigue, shouting for help.

Margaret Borodkin had come to, lying on a large piece of floating debris—perhaps the wall of a house. She was wet and freezing, and her hip and leg ached from where part of the house had fallen on her; she couldn't move. It was several hours after the quake, and darkness was beginning to fall. She'd heard the screaming and crying from the land and had decided to give it a shot to see if she could be heard.

As she related later, whoever heard her voice had called out in return, asking who it was. When she told him, he promised to get help and headed back up the hill.

Although she couldn't move, Borodkin found that she was close enough to the edge of whatever it was she was floating on that she could dip one hand into the water. She tried paddling in this way toward the shore, but it was impossible to gain any headway against all the other debris, and the water was so cold she couldn't stand the pain in her hand. So mostly she just lay in silence out in the cove, hoping that the villagers would return and figure out a way to get her safely to shore.

She wasn't sure how long she had been waiting, but at some point she heard the sound of a boat. It was the *Marpet*, with her brother-in-law George Borodkin and Mark Selanoff aboard, approaching the village. They had ridden out the quake in Whale Bay and had headed back home as fast as they could. Slowly pushing through the debris in the cove, they came upon Margaret, pulled her on board and wrapped her in three sleeping bags.

Up on the hill, Madsen had gotten the battery-powered radio out of the schoolhouse and with some of the villagers had listened to reports of damage coming in from around the state. There were bulletins that Valdez was wiped out, Cordova was heavily damaged, Whittier and Seward were both ablaze. Madsen's mood had shifted: now she had an eerie and panicky feeling. Were they alone in the world? Would anyone know they were still here? Yes, some of their family, friends and neighbors were gone, but there were still forty or fifty souls here atop the hill. Would anyone know to look for them? They would at least have to spend the night here, huddled together on the snow. No one would sleep, that was for certain.

But the *Marpet*'s arrival helped ease the sense of isolation, for the boat had a two-way radio. As soon as they learned of the situation in the village, George Borodkin and Mark Selanoff radioed to Cordova to alert authorities to what had happened and ask for help.

As the night wore on, it became clear that most if not all of the villagers who survived were now together on the hilltop. There were about fifty of them—which meant, by some quick, sad math, that about two dozen Chenegans were missing.

Timmy Selanoff was among the missing, but he was still alive. He had made it fairly high up on the steep hillside he'd been climbing and had seen the destruction of the village. He'd even heard Margaret Borodkin calling from out in the water. But he'd sat there, dazed, as the day grew dark.

After a couple of hours Timmy got up and started walking. He could see the bonfire above the school, so he headed toward that. He was cold—he was walking through snow, and somewhere along the way he'd lost a shoe—and frightened. He began to call out, hoping someone would hear him. Eventually he heard a voice responding from down where the village had been. He followed the voice downhill and found himself in the arms of his father, Charles, the village chief.

Charles embraced his son. "You seen Buttons and Gula?" he asked of Timmy's younger brother and sister. They had been down at the beach playing; Timmy had lost sight of them when he had ventured down to the far end of the cove. It was clear now that they had vanished, taken by the tidal wave.

The screeching was the worst thing. Gloria Day didn't recognize the noise, but later she realized what it was: the sound of her house, just a few blocks from the waterfront in Valdez, struggling to stay in one piece. The sharp shaking of the earthquake was warping the wood frame, so much that studs were bending away from sole plates and joists were being pulled from ceiling beams. The whole house was being bent out of shape, and the steel nails that held the lumber together were taking the strain. The screeching noise Day was hearing was the sound of those nails being pulled through wood fibers.

She and her husband, Walter, ran outside. McKinley Street was rippling and cracking, the waves running from north to south through the property next door and on down the street. Gloria looked down McKinley toward Alaska Avenue and the center of town. The buildings on the corner seemed to rise and fall as they rode the waves. At times they almost disappeared from sight.

Then she turned and looked to her right, toward the waterfront. The stern of the *Chena,* the 441½-foot-long cargo ship that was being unloaded at the main dock, was rising at a sharp angle, its bow pointed down. The stern was so high Day could even see the ship's large brass propeller above the houses.

Down at the dock itself, the world was ending.

When the quake began, the *Chena*'s captain, Merrill Stewart, and its pilot, John Carlson, were eating in the dining room belowdecks. They felt a shaking at first, followed by sharp shocks. The ship seemed to be hopping about. Stewart instinctively knew it was an earthquake, and the two headed toward a ladder that would take them to the bridge, three decks up. Stewart had been aboard ships most of his life, and at age sixty-one he was still pretty nimble. He made it up to the bridge, he estimated later, in about twenty seconds.

Among the crew were a couple of shutterbugs—Ernest Nelson and Fred Newmayer. With little to do during the cargo operation, they had been out on the deck with 8mm film cameras, shooting the scene at the dock to while away the time. With its beautiful snow-draped mountains as a backdrop, Valdez was one of the most photogenic of the ports they visited.

Less than half a minute after the shaking started, that port disappeared, as land turned to liquid. A long slice of the seaward edge of the plain that Valdez sat on—a section nearly a mile long and as much as six hundred feet wide—compacted, slumped and then slid into the bay. When it did, it took the two docks with their warehouses and canneries with it, as pilings and decking and buildings tilted and twisted and finally broke apart and descended into the maelstrom. It nearly took the *Chena.*

In grainy frames from Newmayer's and Nelson's films, the main dock can be seen collapsing and the roof of one of the warehouses starting to fall. A few small boats are in view, looking as if they've been tossed about in turbulent water. But eventually most of the frames are filled with what appears to be a dark, churning mass of water, thick with debris.

The film frames give little idea of the human horror during those moments. The people on the dock who had been working or watching were doomed. Those onboard the *Chena* were witnesses, however, and what they saw they would not soon forget.

From his perch on the bridge, Stewart saw the dock—the dock that his ship was firmly tied to with giant hawsers—start to collapse. It appeared to fold up first, he thought, and was accompanied by a tremendous noise. He saw men, women and children running around in panic, trying to keep their footing as the dock heaved, unsure of what was happening or what to do.

From the deck, Chester Leighton, the ship's chief engineer, saw men running out of one of the warehouses. His initial thought was that something must have exploded inside. Carson Dorney, the second engineer, watched as some of those on the dock ran for safety along the earthen causeway that led to shore, only to be stopped when a huge fissure opened in front of them. As the land they were on collapsed, men tumbled into the water, grasping for anything—timbers, boxes, debris—to stay afloat. One clung to the side of the fissure before he, too, fell in. In the water, some of the victims were caught in a whirlpool of water and debris. Two were sucked down immediately; a third lingered for a while on a floating piece of debris before disappearing.

To Dorney, it was as if the earth were swallowing everyone.

The Stuarts' car disappeared into the water, taking the entire family with it. Jim Growden, the popular high school basketball coach and teacher, vanished with his two preschool-aged sons, Jim Jr. and David, whom he'd brought down to watch the unloading. Dan Boddy, a trucker from Fairbanks who was on the dock to pick up a load from the *Chena,* went down with his truck. Dennis Cunningham and Stanley Knutsen, who had ditched Dan Kendall a short while before, disappeared, and Dorney watched helplessly as a couple of other youngsters who had just gotten some oranges from the cook tried to make it to safety. A man picked up one of them, a young girl who was moving slowly. None of them made it.

In all, in a matter of a few minutes at most, twenty-eight people who had been on the dock were gone. Among them was Jerry Zook, the young man who was to be married in a week.

In town, the earth was fracturing as the shock waves rippled through streets and yards. The biggest crevasse, found north of the Richardson Highway, was six feet wide and four feet deep and ran nearly the length of a football field. As elsewhere in the earthquake zone, the waves rippling through the ground pulled the cracks apart and then pushed them together. In Valdez, with the water table so close to the surface, and with pipes from the town's relatively new sewer system breaking left and right from all the ground movement, the result was a squalid spectacle. Geysers of liquid—a cocktail of water, mud and sewage—spouted into the air, some reaching heights of twenty feet or more. Much of the mixture ended up in the streets, where it was trapped between snowbanks on either side. Residents found themselves trying to drive or walk through pools of mud and sewage up to two feet deep.

Charles Clark, who worked for the state highway department, watched as shock waves roiled the ground and fissures opened and closed near his house. Some opened just a few inches, others as much as three feet. His son was standing some four hundred feet away and rising and falling as the waves passed by. The boy was six feet tall, and Clark used him as a kind of measuring stick: he was fully visible when a crest reached him, but then as he sunk into a trough, his legs disappeared. Clark calculated that the land was rising and falling about three to four feet with every wave that passed through.

In the center of town, drivers were having the same problem they were having elsewhere—cars were bouncing and sliding and not responding at all to steering or braking. More than a few cars ended up with their front or rear wheels in fissures. Utility lines swayed back and forth and the power went out. At Gilson's food store, the owner, George Gilson, and a few customers were inside as the building started to shake and goods crashed off shelves. At the Pinzon and other bars, liquor bottles toppled from shelves and broke. The Glacier Bar in particular was awash in whiskey; the proprietor and two customers had been transfixed during the quake and had watched, unable to react, as bottle after bottle cascaded to the floor.

Despite all the screeching and twisting, the Days' home remained intact, as did most of the other houses and other wood-frame buildings. But many of the commercial buildings in town, made of concrete blocks or other masonry, suffered severe damage. Gilson and his customers experienced this directly: as they made their way to the door, the front wall of the building was heaving and appeared ready to collapse. Nearby, the facade of the Alaskan Hotel *did* collapse as the shaking continued. Other buildings, including Harborview, the state hospital for the mentally disabled out by the highway, suffered severe cracks. In many cases it seemed that where a horizontal ground fissure met a building, the fissure had just continued vertically, up a wall.

At young Gary Minish's home outside of town on the Richardson Highway, most of the family was in the kitchen as his father, Frank, finished the meal he'd decided to have before eventually heading to the dock. When the shaking began he told Gary to stand in the doorway and the others to get under the kitchen table. Frank had built the table himself, out of heavy timbers, and he figured it was one of the safest places to be should the house start to come apart.

From his position in the doorway, Gary could see into the yard. The ground was rippling along, fissures were opening and closing and the trees were flapping back and forth. He was entranced by the power of it all. Yet despite the crazy scene outside, the house itself seemed to be holding together.

Just then they heard a loud noise, a combination of whooshing and splintering, from the living room. Gary looked over. A fissure had apparently opened up directly beneath the room, and then it had closed up again. In doing so it squeezed water, and who knows what else, out and up against the underside of the house. The force of the water had been so strong that it had broken through the floorboards. There was now a geyser of water shooting straight up into the living room. A rug that moments before had been on the floor was now pinned against the ceiling by the water.

Gary and the others watched in amazement as the living room quickly began to fill with water. That was enough for Frank Minish. It was clear now that the house was going to break apart. He ordered everyone out the door. They staggered across the yard and into the family car.

Closer to the center of town, most of Dorothy Moore's family was in the kitchen, getting ready to sit down to a Good Friday dinner of spaghetti. When the quake began, Dorothy, who had been ironing her dress in one of the bedrooms, ran to the kitchen to be with the others. But then she remembered that she'd left the iron on, and what if with all the shaking it fell to the floor and started a fire? She'd never forgive herself if they made it through the quake only to have the house burn down because of her forgetfulness. So she struggled back to the room to turn the iron off. Then she decided it was probably a much better idea to leave the house.

Outside, the ground was shaking so much it was as if it were alive. Moore could hardly believe her eyes. Then she heard the sound of a ship's horn, blaring repeatedly. She knew that a cargo ship had come into port that afternoon, just an hour or two earlier. Now it sounded as if the ship was leaving. But it was far too early for that.

Down at the harbor, the *Chena was* leaving, though not in any controlled fashion. The ship—all 10,812 tons of it—was in deep trouble.

When the waterfront first compacted and slumped and the main dock began to collapse, it forced the *Chena* seaward. Either the ship slipped its mooring lines—not hard to imagine given that the dock and pilings were giving way—or they broke under the strain. So now the *Chena* was adrift, at the mercy of whatever happened next.

When the entire waterfront then slid into the bay, it displaced a huge volume of water, which had to go somewhere. Where it went was up, forming a bubbling mound of liquid directly beneath the *Chena* and lifting it twenty to thirty feet. The ship heeled sharply over on its left side.

Stewart sounded the alarm and ordered the engine room to raise steam. The ship had been in port for only a little more than an hour, so Leighton and the engine-room crew didn't need much time; they got the engine going in about ninety seconds. But until then the ship was without power, and the mound of water was moving toward land as a wave, carrying the *Chena* with it. The ship's bow pitched crazily into the air, then it dipped and the stern rose high, exposing the propeller to Gloria Day and others who happened to be looking toward the waterfront. Then the *Chena* came down onto the water at the spot where the dock had been, hitting with such force that Stewart couldn't comprehend how it stayed in one piece. This was the ship, after all, whose hull had cracked under far less stress more than a decade before, when it was the *Chief Washakie.* "I thought she was done for," Stewart wrote later.

"No ship can withstand that kind of battering." But though the *Chena* shuddered as it thudded into the mud, rocks, broken pilings and other remains of the dock, it stayed intact.

Stewart saw people in the water, flailing about, as the ship was coming down on top of them. The image would keep him up at night for weeks afterward.

The *Chena* was now on the back side of the wave, rolling sharply on its right side and still heading toward the town. Forest L. Sturgis, an engineer with the state highway department, watched the unfolding scene from his apartment in the Valdez Hotel on Alaska Avenue. To him it looked as if the ship were jet propelled, it was riding so fast on the incoming water.

Nine Valdez men had been working on the ship itself—two on deck and seven in the holds, putting the cargo on pallets so that it could be lifted out. The steep rolling and other movements of the *Chena* during these first moments of the quake spelled doom for some of them.

Down in hold no. 3, barrels of asphalt broke loose and rolled into Howard Krieger, killing him. Almost simultaneously a heavy steel hold cover broke and fell from above, first striking and killing Paul Gregorieff, who was the foreman of the crew in the hold. The cover then hit Jack King, crushing his chest and legs. He survived but had to have both feet amputated.

The *Chena* was still at the mercy of the roiling water, which had moved into town, two blocks inland from McKinley Avenue. The ship was pushed farther toward the shore, between what remained of the two earthen causeways, where the small boat harbor was. The water had already wreaked havoc on the harbor, swamping most of the seventy boats and sweeping them from their berths. For a brief moment, the harbor was empty; the wave had passed inland and the *Chena* was high and dry. But soon the water started flowing *out* of town, the backwash from the wave. By now the *Chena* had regained power and was floating again—or at least part of it was. Its stern was stuck in the mud, pilings and other debris that remained from the main dock.

The outflow, though, proved a lifesaver. It gently pushed the bow of the ship away from the shore, enough so that the *Chena,* its propeller now slowly churning through the debris, was able to push past the remains of the smaller dock. The ship kept slowly moving south, hugging the shore, trying to get enough speed to overcome the water that was forcing it inland.

Aiming his camera back toward town, Nelson recorded an astonishing sight. Where the docks had been there was now a cliff. Water flowing back from the town was cascading over it. It was as if, in the few minutes of the quake, a gaping chasm and waterfall had formed. No one had ever seen anything like it.

After a few more minutes of wallowing in the mud near the shore, the *Chena* was finally able to turn into the bay. The water intakes for the steam engine were now clogged with debris and mud, and the engine was in danger of overheating; Stewart had it shut down. He then ordered Neal Larsen, the chief mate, to do a quick damage survey. After a few minutes, Larsen reported that the ship was not taking on water; it appeared to have survived the rough ride. Stewart, who had had the lifeboats manned and ready to go at a moment's notice, couldn't believe his good fortune.

The *Chena* raised steam once again and headed into the bay to take stock of the dead and injured. They badly needed a doctor for King and for Ralph Thompson, a crew member who, amid the panic, appeared to have suffered a heart attack. Perhaps the town doctor could be brought out if a skiff could be found. And there were surviving longshoremen from Valdez who needed to get back and check on their loved ones.

A little while later, as the *Chena* cruised slowly along, the crew saw something astonishing: there were dead fish everywhere, completely covering the surface. They appeared to be red snappers.

10

STUNNED

The tire is about three feet in diameter and is mounted on a steel rim. It's a heavy-duty one, built of ten rubberized layers, off a forklift truck. In the photograph, taken a few days after the earthquake at what remained of Two Brothers Lumber Company, a sawmill on the waterfront in Whittier, an unidentified man is standing behind it. With his left hand he is keeping the tire balanced upright. With his right he is gesturing, showing what happened to it. There's no need for that, really, because what happened to the tire is plain to see, if scarcely believable. A two-inch-by-eight-inch plank is sticking through it, the lumber having gone clear into one sidewall and out the other. It's as if the tire had been pierced by a giant, and very blunt, arrow.

In the hours after the earthquake, dazed Alaskans tried to take stock of what had hit them. One thing was obvious: this was an earthquake with a power far greater than anyone in the state had experienced before. They would eventually learn that the Good Friday quake, as everyone took to calling it, had a power far greater than just about anyone *anywhere* had experienced before. No, an atomic bomb hadn't hit the state, as some had initially thought, but by one quick calculation that was published in the newspapers in the days following the quake, the energy released was equivalent to thousands of A-bombs.

The speared tire was just one of many artifacts that testified to the forces that had been unleashed—in this case, in the form of a thirty-foot wave that had struck the Whittier waterfront and sent a barge loaded with lumber careening into the sawmill. There were so many others: hemlocks that had been snapped in half at midtrunk height; spruces that were split vertically as the ground cracked beneath them; heavy equipment that had been moved hundreds of feet; hillsides that had been stripped bare of vegetation by the scouring action of water; one-ton boulders found halfway up slopes; roads split down the middle and sunk ten feet or more on one side; rail lines in tangles; brick building facades in jumbled heaps; cars upside down; *homes* upside down; boats littering streets. Some of the "artifacts" were enormous. Near Sherman Glacier, about twenty miles east of Cordova, the top 500 feet of a 4,300-foot-high mountain (later named Shattered Peak) broke away during the quake. In the ensuing landslide, an estimated twenty-five million cubic yards of rock hurtled down the mountain. With a thin cushion of air trapped beneath it that reduced friction, the rock debris reached speeds of two hundred miles an hour and traveled more than three miles. It eventually spread out over about five square miles of the glacier in a layer four feet thick, like the frosting on a cake, where it sits to this day.

Stopped clocks showed that the quake had begun at 5:36 p.m. By most accounts, it had had a beginning, a middle and an end—starting slowly, quickly going into overdrive with violent rolling and jarring and then mercifully fading out. Estimates of the duration varied widely. A fisherman in Prince William Sound thought it lasted only a minute or two, while to at least one person, trapped inside an Anchorage store, it seemed that the nightmare had gone on for ten minutes or longer. The consensus, however, was that the shaking had lasted four to five minutes. Most agreed that, no matter the exact amount of time, it felt like forever. The earthquake had been unbearable.

Scientists immediately tried to quantify the quake's power, to put a number on what had scared Alaskans out of their wits. Initial reports were that it registered 8.2 to 8.7 on the Richter scale, putting it on a par with or stronger than the 1906 San Francisco earthquake, at least in terms of the energy released. The lowest number came from the seismological laboratory at the California Institute of Technology in Pasadena, from none other than Charles F. Richter himself, who with Beno Gutenberg had developed the scale that bore his name in the 1930s as a way of expressing the relative power of earthquakes. (In the 1950s, Richter had done some backward reasoning on the San Francisco quake and come up with 8.25 as its magnitude.)

The Alaska earthquake had been centered, seismologists said, at a spot fifteen miles below the Chugach Mountains, near a long finger of Prince William Sound called Unakwik Inlet. Anchorage was seventy-five miles to the west and Valdez fifty-five to the east.

It had made the earth ring like a bell, as one scientist later put it. Another likened the planet's response to that of a struck tuning fork. The quake was picked up by seismographs all over the world, including about one hundred that had been installed since 1960 at government agencies and universities as part of what was called the World-Wide Standardized Seismograph Network. The network was funded by the US military's research wing, DARPA, and its main purpose was to detect nuclear weapons tests, which produce seismic signals that are different from those of earthquakes. (Not surprisingly, the Soviet Union and its satellite states were not involved.) But the program had many benefits for peaceful science as well. The one working seismograph in Alaska at the time, in Fairbanks, was part of the network but had been overwhelmed by the power and proximity of the quake.

Other instruments detected the event as well. As the shock waves traveled through the earth they caused groundwater levels to fluctuate. In the United States, gauges at more than seven hundred wells—including some as far away as Georgia, Florida and Puerto Rico—recorded changes. At an astounding twenty-three feet, a well in Belle Fourche, South Dakota, near the Wyoming line, had the biggest fluctuation of all. Outside of the United States, wells were affected in England, Belgium, Libya, Israel, South-West

Africa (now Namibia) and Australia, among other countries. Rivers, streams, reservoirs and lakes showed temporary changes in level, too, at more than 750 surface-water gauging stations around the United States. Even the Gulf of Mexico was affected: the seismic waves caused seiches, oscillations in the water level, like water sloshing around in a bathtub. Some of the resulting waves were six feet high and caused minor damage to boats and docks.

The earthquake *was* like a nuclear explosion in one respect: it had an effect on the atmosphere. The rapid land movement over such a huge area compressed the air above Alaska for a moment, causing a brief spike in air pressure that spread out in waves, like the ripples on a pond. Sensitive barometers at the Scripps Institution of Oceanography near San Diego, nearly 2,500 miles from Anchorage, measured the slight pressure changes about three hours after the quake.

Earthquakes are always followed by aftershocks, and this one was no exception. In the first twenty-four hours, seismographs recorded twenty-eight of them, including ten that at magnitude 6.0 or greater would qualify as strong earthquakes in their own right. By two months after Good Friday, some twelve thousand aftershocks greater than magnitude 3.5—meaning, generally, those that can be felt—had occurred. Aftershocks become less frequent and weaker over time, and that was the case with this quake. Still, over the next year and a half, there were thousands more small tremors before the region finally calmed down.

In the first days after the quake, Alaskan officials began to tally the cost, both in lives and in property. Civil defense, Red Cross and Salvation Army representatives came up with initial estimates of one hundred or more dead and more than 450 injured. Federal emergency planning officials made an equally quick estimate of the cost to clean up and rebuild: up to half a billion dollars.

If those figures seem low, remember that Alaska's population at the time was only a quarter of a million and that its yearly gross state product—the value of all the goods and services produced—was only about $1 billion. If the quake had struck a more thickly settled region, or a more prosperous one, the toll in lives and property would have been vastly higher. Likewise, if it had struck earlier on a busy weekday instead of late on a holiday—when more people would have been out and about, at work or school or shopping—and at high tide rather than low tide, no doubt there would have been more death and destruction. In some ways, Alaska was lucky.

While the quake was felt over a wide arc, major damage was confined to a lesser one, from Kodiak Island in the southwest up through Anchorage and southeast through Cordova, encompassing much of the Gulf of Alaska and all of Prince William Sound. This area included most of what would be considered "developed" Alaska. Most of the state's infrastructure was here, and much of it was now in tatters. Valdez was not the only port to be effectively destroyed. At Seward, where cargo from the Lower 48 was unloaded (and where the *Chena* had stopped before Valdez), docks and warehouses were wiped out and oil tanks were in flames. At Whittier, Two Brothers Lumber was only one of many businesses that were gone, and the town's oil tanks, too, had caught fire. The state's fishing industry had been hit hard, with canneries and dock facilities heavily damaged or obliterated by high water around the sound and the Gulf of Alaska. The fishing port of Kodiak had also been swamped by tidal waves, and much of its fleet was now in its streets. Cordova suffered little direct damage, but a Coast Guard ship that arrived a day after the quake found changes: suddenly the harbor was not nearly as deep as before, and dredging would be needed before the town's fishing fleet could operate efficiently there.

Roads throughout the region were impassable, including the vital two-lane highway linking Seward and Anchorage. As George Plafker had seen in that first reconnaissance flight, it was blocked by landslides and avalanches, and many bridges had collapsed. The spur road from Portage to Whittier was in bad shape, too, although the 2.5-mile-long tunnel just on the landward side of Whittier, burrowed through Maynard Mountain by the army in the 1940s, appeared to have suffered no ill effects from the shaking.

Even more devastating than the damaged roads, however, was the condition of the Alaska Railroad. South of Anchorage, the 470-mile-long line, the spine that tied the state together economically by bringing goods as far north as Fairbanks, was wrecked. Early estimates were that repairs would take months.

Other links were cut. A section of the Million Dollar Bridge in the Copper River delta, which had been built for the railroad in the early 1900s to reach rich copper deposits (at the then unheard-of cost of nearly $1.5 million) but now served motorists, had been shaken off its foundations. Repairing it would be a gargantuan task. Anchorage's international airport was out of commission. Other airfields had cracked runways.

The military had its own problems. There were concerns about the Cold War communications infrastructure—that was one of the reasons the officers had been happy to see Plafker and the other geologists arrive the day after the quake. But commanders kept quiet about a more immediate and serious worry. Some of the army's Nike Hercules missile batteries, which were designed to shoot Russian bombers out of the sky in the event of World War III, were damaged. There were three batteries around Anchorage, manned by crews from Fort Richardson, and two of them had suffered from the earthquake shocks. At one, at Fort Richardson itself, a large radar dish had shifted off its pedestal, and the army was looking at repairs that might take months.

But far more ominous was what happened at a battery just off the end of the main runway of the Anchorage airport. Some of the forty-foot-long missiles, stored in aboveground bunkers, had toppled during the quake, cracking open and spilling highly flammable solid rocket fuel. The danger of an explosion and fire was high, and adding to the risk was a characteristic of the missiles that the army didn't talk about: they carried nuclear warheads.

Although the warhead on each missile was about one-third more powerful than the bomb that had destroyed Hiroshima, there wasn't much danger of Anchorage going up in a nuclear fireball. But the risk of radioactive contamination—of a conventional explosion that could send some of the warhead's radioactive plutonium into the air, to eventually settle across the city and elsewhere—was very real. The military immediately took care of the problem, sending in a crew of soldiers who worked virtually nonstop over three days and amid the occasional aftershock to delicately clean up the mess.

But civilian Alaska needed help, too, and fast. The army and air force were mobilized. Personnel and supplies were flown in from bases in the Lower 48, and relief flights fanned out to the areas where damage had been reported.

In Anchorage, where power was out, troops roamed a thirty-square-block area as rescue and recovery efforts got under way. Fire department and other rescue units started a house-by-house search in Turnagain Heights, with victims being taken away in ambulances and private cars. Military helicopters swooped overhead, trying to spot survivors amid the jumbled blocks of earth. Downtown was a disaster, with the devastation on Fourth Avenue and the rubble of the facade of the J. C. Penney store littering Fifth Avenue. Search-and-rescue teams went through every building, but the Penney store was so big, and its damage so severe, that it required a dedicated team of searchers.

One elementary school was a total loss, and a hospital for the native population narrowly escaped destruction when land nearby gave way. In addition to the unfinished apartment building that collapsed, two identical, and occupied, fourteen-story residential towers were damaged. A large hotel, the Westward, just a few blocks from Fourth Avenue, seemed fine from the exterior, but the internal structure was severely damaged. In all three buildings, amazingly, no one was injured.

Amid the devastation there was the occasional odd, calming sight, like the small clothing store where the display windows had shattered and a group of mannequins now leaned out onto the street, or the florist's shop with the fresh flowers still sitting in their vases. The Denali movie theater, despite having dropped about ten feet when Fourth Avenue cracked and slid, had not suffered a single broken lightbulb on its marquee.

In the slide at Turnagain Heights—the one Bob Atwood lived through—about seventy-five of the neighborhood's homes were destroyed. The scale of the slide was enormous: two hundred acres of land had moved as much as two thousand feet toward the water. Even some land *under* the water had moved. But Turnagain was only the largest of several slides in Anchorage. The damage on Fourth Avenue downtown was due to a smaller one, and the school that was destroyed had been brought down, literally, by a slide on Government Hill, on the other side of Ship Creek. The school had split in two, and one half had dropped a dozen feet below the other.

Most of the damage from the quake, in Anchorage and elsewhere, was caused, not directly by shaking, but rather indirectly by the effects the shaking had on soil and sediments. The destruction of the waterfront at Valdez was the most obvious example, as the sediments that the dock sat on turned to jelly and slumped during the quake. Seward appeared to suffer roughly the same fate, with a big chunk of the waterfront sloughing into the water. The Anchorage slides were evidence of movement as well, but the action was a little different. The sliding occurred in a layer of fine and slick sediments known as Bootlegger Cove Clay. The soil above was carried along by this sliding clay layer, which is why at Turnagain Heights it broke into large blocks that moved about crazily and nearly killed Atwood.

No one was certain yet, but the waves that had hit Chenega appeared to have been caused by an underwater landslide of nearby sediments deposited by glaciers. That would explain why the first wave arrived while the ground was still shaking. Whittier appeared to have been hit by such a locally generated tidal wave as well.

But there were other, much farther-reaching tidal waves spawned not by landslides but by the ground movement along the fault—in other words, by the earthquake itself. (In their circular published a month after the earthquake, Plafker and his colleagues referred to these as "seismic sea waves.") These traveled far across the Pacific at hundreds of miles an hour, even reaching the Antarctic Peninsula, ten thousand miles away.

Alfred Wegener, who developed the theory of continental drift, during an expedition to the Greenland ice sheet in 1912–13. *Courtesy of the Alfred Wegener Institute*

Harry Hess, a Princeton geologist who came up with the idea of seafloor spreading, made measurements of the ocean floor while on active duty in the navy during World War II. *The Department of Geosciences, Princeton University*

George Plafker at a spike camp in Alaska in the 1960s. *Courtesy of George Plafker*

The Don J. Miller, the barge that George Plafker and others used for their work in the summer of 1964. *US Geological Survey*

One of George Plafker's assistants holding a rod next to barnacles to measure elevation changes from the earthquake. *US Geological Survey*

The damage on Fourth Avenue in Anchorage, where half the street dropped about ten feet. *US Geological Survey*

The tire that was pierced by a plank in Whittier during the earthquake. *US Geological Survey*

Destroyed houses in the Turnagain Heights neighborhood in Anchorage. *US Geological Survey*

Debris covered Sherman Glacier after a landslide that occurred when the peak of a nearby mountain broke off during the earthquake. *US Geological Survey*

An area of the seafloor at Montague Island that rose up during the quake. The white color is from marine organisms that died when they were exposed to the air.
US Geological Survey

The village of Chenega, including the schoolhouse on the hill.
Alaska State Library

Village life in Chenega centered around the Russian Orthodox church. *Courtesy of Kris Van Winkle*

Kris Madsen and her students put on a Christmas pageant in 1963, with the help of the lay priest Steve Vlasoff. *Courtesy of Kris Van Winkle*

Survivors gathered on the hill above the schoolhouse after the quake. *Courtesy of Kris Van Winkle*

Valdez from the air, looking toward the town docks, in 1962. *Bob and Marie Logan slides, Archives and Special Collections, Consortium Library, University of Alaska Anchorage*

Little was left of the village homes and the church after the earthquake. *Courtesy of Kris Van Winkle*

Valdez several days after the earthquake. At the bottom, the town docks are gone. *National Oceanic and Atmospheric Administration/Department of Commerce*

George Plafker on Montague Island in Prince William Sound in 2015. *Courtesy of the author*

When a tidal wave is out in the deep ocean, it has little or no effect on the surface. Such a wave can pass under ships without their crews realizing it. But when it approaches land, local conditions—the topography of an inlet, say, or the contours of the bottom of a bay—can have an enormous impact, focusing the wave energy. Waves can grow taller and can slow to a tenth of their initial speed or slower.

The first wave of this type had struck Kodiak Island within an hour, as more of a gentle flood than a wave. But a second wave, about an hour later, was a thirty-foot wall of water that rushed into the harbor, picking up fishing boats like toys and pushing them five blocks inland. The torrent wiped out all the docks save one, and as it raged into the town close to sixty buildings were destroyed. The Kodiak Naval Station nearby was flooded and heavily damaged—its cargo dock, its roads and bridges, its central power plant, its radar installation and even the station's bowling alley were destroyed. In all, nineteen people died on Kodiak, including eight in the town itself.

Meanwhile, waves were spreading out across the Pacific. The first one reached the Hawaiian Islands, some 2,700 miles to the southwest, about five hours after the quake, and Japan a few hours after that. It took almost twenty-four hours to hit Antarctica. In most cases where they struck land the waves weren't very big and did little appreciable damage. Hilo, Hawaii, had some of the biggest waves, at about ten feet. Waimea, on the North Shore of Oahu, had fifteen-foot waves, but the famous surfing spot regularly has forty- or fifty-footers that roll in during the winter. In Japan, the waves from the Alaska quake were much smaller, a foot or less. Those that reached the West Antarctic Peninsula were about four feet high.

But larger waves—some more than twenty feet high—left a trail of destruction southeastward along the North American coast, through British Columbia, Washington, Oregon and California and into Baja California, in Mexico.

Cape St. Elias, at the tip of Kayak Island, sixty miles southeast of Cordova, was hit first. Four Coast Guardsmen lived at the cape, maintaining a lighthouse that had stood there, in one form or another, since 1914. One of the guardsmen, Frank Reid, had been out photographing wildlife when the quake occurred, and had had his leg broken by falling rock. His three colleagues came to find him and, while carrying him back, were hit by the wave. The three able-bodied men managed to swim to safety, but Reid was lost.

The wave continued southeast along the Alaskan and then Canadian coasts. Vancouver Island, in southwestern British Columbia, first encountered it about three hours after the quake. The most devastation was caused at the twin towns of Alberni and Port Alberni, in the southern part of the island and up a twenty-five-mile inlet from the Pacific. Because it took time for the wave to travel up the inlet, the towns were not hit until about midnight. That first wave served, literally, as a wake-up call. Residents knew that a first wave was usually followed by others, so after the first one townspeople went around rousing people and urging them to get to higher ground. When the second wave, estimated at about ten feet, struck an hour later, it washed away fifty-eight homes and stores and caused damage estimated at $10 million. But there was no loss of life.

Farther south, waves struck one stretch of the coast after another. At the southern edge of the Olympic Peninsula in Washington State, high water washed away a bridge in the town of Copalis Beach. In Cannon Beach, a sleepy town in Oregon, participants in a late-night poker game ignored one telephone call warning them of the possibility of a wave; a second call a short time later relayed word that a wave had just hit the shoreline. Some houses floated away and a bridge into the town was destroyed, but everyone, including the poker players, survived.

Just one hundred miles down the coast, however, at Beverly Beach, near Newport, Oregon, the McKenzie family was not so lucky. Monte McKenzie, a Boeing engineer, had driven down from the family's home in Tacoma, Washington, with his wife, Rita, and their four children, ages three to eight, for the Easter weekend. The family had been through hard times recently—their eldest child, ten-year-old Susanne, had died from burns suffered at a campfire seven months before. At Beverly Beach the McKenzies had come across a driftwood shelter and had gotten permission to sleep there. Around 11 p.m. they were awakened when a small wave washed up the beach, reaching their campsite. The couple grabbed the children and started making their way inland, but were soon overcome by a series of much higher waves. A strong swimmer who had taught all her children to swim, Rita McKenzie tried to hold on to two of her children. The four youngsters were swept away and only one body was ever recovered—that of their six-year-old, Ricky. "I have no idea what happened," Rita McKenzie said later, describing the tragedy. "Nobody had a chance."

Waves wreaked havoc farther down the coastline. Boats were sunk in Santa Cruz and Half Moon Bay, south of San Francisco. A ten-foot-high wave hit Catalina Island in Southern California. Water surged into Marina del Rey, the newly built community just north of Los Angeles International Airport, and tore up 450 feet of dock and pushed it a half mile inland. In San Francisco, warnings on the late news about the possibility of a tidal wave drew an estimated ten thousand people to the shore, curious to watch the spectacle. Fortunately the waves that began coming in after midnight were not very high.

But it was in Crescent City, California, a small port in timber country ten miles south of the Oregon border, that the waves proved most lethal. The death toll remains the highest from a tidal wave in the continental United States.

Crescent City, a town of three thousand and the seat of Del Norte County, had experienced tidal waves before. Coastal engineers eventually determined that several factors, including the shape of the town's harbor, tended to focus the wave energy and make waves worse there than just about anywhere else on the West Coast.

News of the quake in Alaska had reached the town, and a tidal wave warning had been issued shortly after 11:00 p.m. Efforts were made to evacuate coastal and low-lying areas. Not everyone got out, but the first two waves that hit the town, beginning at 11:52 p.m., were small. They pushed into the harbor and Elk Creek, causing minor flooding in the commercial district.

This is where the town's familiarity with tidal waves proved fatal. Previous events had been minor, with one or two waves that did not do much damage. After the second wave on this night, many Crescent City residents thought this one would follow the same pattern. They returned to their homes and businesses to clean up.

At the Long Branch Tavern, on Highway 101 near Elk Creek, the owners, Bill and Gay Clawson, came back to the bar with two employees and their twenty-nine-year-old son, Gary, and his fiancée, Joanie Fields. They hadn't closed down properly and hadn't collected the night's proceeds from the till. Once at the bar, the water seemed to have calmed, and it was Bill Clawson's birthday, so they decided to celebrate with a few beers. Two other friends stopped by to share in the merriment.

Shortly after that, the third wave struck. Later, residue on a flagpole near the harbormaster's office showed how high the water reached: nearly twenty-one feet above mean sea level. The wave flooded the tavern and pushed it off its foundation. For the eight people inside, the only way to go was up—first to the second floor, then, as the water kept rising, to the roof. Gary Clawson and one of the latecomers were the only ones who knew how to swim, so they jumped in the water and swam and then walked to higher ground. From there, Gary got a small rowboat and rowed back to get the others. With seven people on board the boat was difficult to maneuver, but they got it headed in a direction that they hoped would lead to land. At that point, though, the water started receding, about as fast as it had come in. The boat was drawn into Elk Creek by the strong current and eventually smashed into a bridge. Five of the seven occupants, including Gary's parents and his fiancée, drowned. Gary was one of two who survived.

Five other Crescent City residents died as a result of this third wave or a fourth one, equally strong, that arrived a little later. Nearly three dozen people were injured. A thirty-block area in the center of town, containing 250 homes and businesses, was essentially destroyed. Adding to the disaster, a loaded gasoline tanker truck got picked up by the third wave and was slammed into a Pontiac dealership, starting a fire that destroyed the dealership and spread to nearby oil-storage tanks. The blaze burned for three days.

In all, sixteen people died as a result of the waves in California and Oregon, including one in Klamath, California, not far from Crescent City, and another in Bolinas Bay, north of San Francisco. In Alaska, the death toll remained uncertain for weeks, which is not surprising given the remote character of much of the state and the sudden disappearance of many of the victims. Most calculations today put the state toll at 115; adding the California and Oregon deaths brings the overall tally to 131.

One thing quickly became clear about the Alaska fatalities, however: relatively few people were killed in the way one would think that most deaths occurred during earthquakes, by being struck or crushed by falling structures or debris. Statewide only twelve fatalities seemed to fit this type, of which nine were in Anchorage. Of those, five occurred downtown, including two around the J. C. Penney store. (Blanche Clark, however, who had been trapped in her crushed Chevrolet, was not one of them. She was cut out of the car and, despite suffering a broken neck and arm, cracked ribs and a punctured lung, made a complete recovery.) Three people were killed in Turnagain Heights, including Leora Knight, a high school science teacher who fell into a crevasse along with her husband, Virgil. The crevasse soon closed up, crushing them. Leora never regained consciousness; Virgil survived but had to have a leg amputated.

Most of the Alaska deaths—103 by modern count, out of the 115—were caused by waves. In addition to the 19 on Kodiak Island, 12 died in Whittier and 11 in Seward. There were isolated deaths elsewhere in Prince William Sound and the Gulf of Alaska.

Then there were Chenega and Valdez, where the loss of life was worse than anywhere else. In the hours and days after the quake, the people of those two communities struggled to cope with the deaths of family and friends, and with what to make of their own lives from there on.

The buzzing sound started up the next morning, after a night short on sleep and long on misery for the sorry group of survivors gathered on the hill above what had been the village of Chenega. The noise came from the sky, faint at first but growing louder. It was the sound of an airplane, a hopeful sign, evidence that the villagers weren't alone and that help would be arriving.

There were actually two aircraft nearing Chenega Island at that moment, around 9 a.m. One was a Coast Guard plane making a broad sweep of Prince William Sound, searching for people who might need help. The other was a seaplane belonging to Cordova Airlines and flown by one of the airline's pilots, Jim Osborne. He was accompanied by Frank Eaton, a member of the Civil Air Patrol. They were on a much more specific mission.

Osborne knew Chenega and its inhabitants well. Cordova Airlines had the mail contract for most of the sound, so Osborne stopped at the village as often as twice a week in summer and every other week in the winter, if the weather allowed. He took the business of delivering Chenega's mail seriously, as he knew it was the main connection most of the villagers had with the world. At the airline office late Friday evening after the earthquake, Osborne had been alarmed to hear reports of the disaster relayed from the *Marpet.*

Osborne decided to fly to the island the next morning, with the idea of shuttling back survivors. For the task he'd need the airline's Grumman Widgeon, a six-passenger high-winged plane that could take off and land on wheels or on its belly on water. But there was a problem. The Widgeon was out of service for maintenance and inspection. Some parts of the plane had even been disassembled.

Osborne called the airline's mechanics, described the situation and asked if they could get the aircraft back in one piece by morning. They were reluctant at first, as the plane was down near the water and they were worried about being caught by more destructive waves. But eventually they agreed. Working overnight, the mechanics had the Widgeon back together before 8 a.m. A little while later, Osborne and Eaton were airborne for Chenega, less than one hundred miles to the west.

As they got closer to the village, flying over Knight Island Passage along Chenega Island's eastern shore, they were overcome by what they saw. The tide had swept most of the ruins of the village out of the cove and into the passage, where they now covered a huge expanse of water—perhaps five square miles, Osborne thought. There was debris everywhere, flotsam and jetsam that had once been Chenega interspersed with dead red snappers, the fishes' skin giving an eerie iridescent sheen to the surface. Much of the debris was unrecognizable, but looking closer Osborne and Eaton could make out intact walls and other parts of houses.

They saw the Coast Guard plane pass overhead as they swooped down toward the village. The pilot made no sign—no dipping of the wings or other maneuver—that he recognized there was a problem below. Osborne didn't give it much thought as he landed in the cove and taxied up to the beach. He had flown in here often, and this time it felt like there was something different about the shoreline; the beach seemed much wider. Osborne wondered if they had arrived at an exceptionally low tide, or if somehow the earthquake had caused a change in the land.

As the plane pulled into the shore past where the dock used to be, some of the village men came down the hill from the schoolhouse. They seemed stunned and emotionless, Osborne recalled later. He could tell from the looks on their faces that they'd been through hell.

Later, Osborne learned that the pilot of the Coast Guard plane had radioed in that the village appeared to have survived the quake. From the air he'd seen the intact schoolhouse and the empty expanse at the bottom of the hill, next to the beach. The waves had done their job so well, and the pilot was so unfamiliar with the village, that he hadn't realized that he was looking at a scene of destruction, a place where fewer than twenty-four hours before there had been homes and a church. Still, Osborne couldn't understand why the pilot hadn't recognized that something was amiss. Hadn't he seen the miles of debris on the water?

The villagers told Osborne that some of them had come down the hill earlier that morning, at daybreak, in the vain hope of finding survivors. They had found a baby bottle filled with milk, which they brought back up to be shared by the youngest children. Then they had discovered the body of Tommy Selanoff, one of two twin toddlers who were lost, in a tree where the water had carried him. Down on the beach they found another body, that of Anna Vlasoff. Her husband, Steve, the lay priest, had been off the island on Friday, in Cordova. Eventually he would conduct funerals for all those who died. The first, the following week, would be for his wife.

Also early that morning, Mickey and Nick Eleshansky had come back from Prince of Wales Passage in their fishing boat, the *Shamrock.* Chenegans now knew the full toll of the disaster that had befallen their village a little more than twelve hours before. It appeared that twenty-three villagers had died; there were fifty-three survivors.

As Kris Madsen came to realize later, the waves had mostly claimed the young and the old. There were reports that a few of the older villagers had instinctively run to the church, rather than up the hill, when they heard shouts about a tidal wave. But others had tried, and failed, to get high enough up the hill to escape the torrent. Some of the youngest victims never made it off the beach where they had been playing.

Twelve children, ranging in age from one to nine, were among the dead or missing. They included Rhonda Eleshansky, just a few weeks shy of one year old, the youngest victim; Robert Selanoff, who, like his twin, Tommy, was not quite two, and Avis and Joe Kompkoff's daughter Jo Ann, who had died along with Willie and Sally Evanoff, Avis's adoptive parents. Madsen counted two victims among her students—Julia Kompkoff, age nine, who had been swept away while trying to flee alongside her father, Nick; and eight-year-old Cindy Jackson, who had perished with her four-year-old brother, Dan, and their mother, Dora.

Anna Vlasoff, the oldest victim, had turned sixty-nine the previous August. Only two of the dead—Dora Jackson and Richard Kompkoff, Avis's cousin—were in their twenties. Several villagers said they had seen Richard swallowed up by the water as he was imploring Anna Vlasoff, who was frozen in fear, to flee to higher ground.

Meeting with the men down on the beach, Osborne planned the evacuation. He'd first take any injured, along with pregnant women and small children. There were a few mattresses in Madsen's quarters in the schoolhouse; these were brought down to line the floor of the plane. A couple of cots in the school were used as stretchers to bring down Margaret Borodkin and Dorothy Eleshansky, who had a deep gash on her back.

The two women were put on the plane, and were joined by as many other women and children as Osborne thought the Widgeon would bear. With Eaton staying behind, Osborne taxied off the beach and out onto the water. Before long the first of Chenega's survivors were heading to Cordova.

Osborne made three flights that day, taking all of the women and children and a few of the men. Madsen was on the last flight, and like the others in the village with dogs, had to leave Tlo behind. The rest of the men sailed back to Cordova aboard the *Shamrock* and the *Marpet*, arriving the next day.

On Sunday the villagers got more bad news. Three people who had lived in Chenega but now made their home at Port Nellie Juan were presumed dead as well: Alex Chimovisky—godfather to Nick Kompkoff—his wife, Anna, and their adult son, Emmanuel, who served as winter caretakers at the Nellie Juan cannery that Chenegans supplied with salmon during the summers. A Coast Guard cutter had been through the area on Saturday and had discovered that the dock had been destroyed by a wave. Two skiffs were found, overturned, with their motors still running.

In Cordova, Reverend Bert Hall, pastor at the Community Baptist Church and also the local Red Cross representative, heard of the destruction of Chenega and developed a plan. Hall and other leaders in the town of 1,500 had had recent experience in disaster management. A little less than a year before, much of Cordova had burned in a fire. In all, fourteen buildings had been destroyed and nearly 10 percent of the population made homeless. Hall and the others had sprung into action, sheltering many residents in the social hall in the church basement until new housing could be found.

Hall now planned to provide similar assistance to the Chenegans. With the hall's basketball hoops and gymnastic rings tied out of the way, and with large curtains strung around the room to give families some privacy, forty-two Chenegans moved in and made it their home. Hall also provided the use of his parsonage for other villagers. A cannery lent cots, and towels and blankets were provided by Cordova's hospital. Local stores provided food that the villagers cooked for themselves using the hall kitchen.

The villagers were appreciative of the help, but for a people who were used to living close to nature, spending much of their time outdoors hunting, fishing and playing, Cordova was a shock. The community hall was crowded, and there was little to do. The Chenegans grieved for the loss of their loved ones, to be sure, but of so much else as well: their homes, their island, their way of life. They were understandably numb and withdrawn.

In the middle of that first week, a Red Cross volunteer from San Francisco, Claude Ashen, arrived in Cordova to help. While at first glance it didn't seem as if he had much in common with the villagers, he would prove instrumental in helping them get back on their feet.

Ashen, known as Pete, was an insurance salesman with John Hancock who in his spare time helped direct disaster-relief efforts for the Golden Gate Chapter of the Red Cross. The night of the earthquake, Ashen received a call from someone he didn't know in

Fairbanks. Over a scratchy connection the caller told him about the disaster and, before the line went dead, begged him to send help to the state. Ashen called his higher-ups in the Red Cross, who met the next day and decided to send him to Alaska.

Ashen had been in Anchorage for several days when, upon hearing reports about the disaster at Chenega, he realized he could be more help to the surviving villagers. Over the next eight weeks, with funds provided by the Red Cross chapter back home, he worked to replace some of what the villagers had lost—boats and gear—so that they could resume their hunting and fishing lifestyle.

Kris Madsen found herself largely on her own. She was lodged separately in Cordova with a local teacher. For her the earthquake had been a figurative as well as a literal jolt. It had helped crystallize her thinking about her future.

State education officials and the Bureau of Indian Affairs had a plan to get the children of Chenega back in classes, making use of a single room at the local elementary school in Cordova. Madsen had been in Cordova about a week when they came to her with a request. A chartered boat was taking various officials and villagers back to the island, and they needed the attendance and other records of Madsen's pupils. Would she go back and get them?

Madsen reluctantly agreed, but she was upset at the request. Chenegans had been through a horrific disaster; villagers had suffered the loss of family and friends and everything they owned. Madsen had had it easier than the villagers themselves—she hadn't lost any loved ones and didn't have much property to lose. But she was deeply traumatized. And all the bureaucrats seemed to care about was paperwork.

Nonetheless she had gone back, along with the chief, Charles Selanoff, and others. Where the center of the village had been, little was recognizable, just fragments of wood and other materials scattered about, with the log pilings that had supported the homes poking up through a layer of silt. When she walked up the hill to the schoolhouse, she found the generator shack and the play yard awash in silt as well. In the schoolhouse she gathered a few things from her living quarters, although most of her clothes and shoes had been given out to the villagers the night of the quake. Then she entered the classroom. It was a shock to be there again, in a place that had once been filled with happy children. The room was as they had left it when Jim Osborne had come to get them, with books, papers, food and other supplies strewn on the floor. Paper Easter eggs still adorned the walls, and the chalkboard still carried the day and date of the earthquake, written in her hand. She gathered up the necessary papers and left.

A few objects were found and salvaged during that trip, including a Russian-language Bible from the church, a basket made by Katy Selanoff, wife of the chief, and some photographs. But the visit was unbearably sad. Officials shot the dogs that had remained in the village, including her dog, Tlo. But Madsen realized that everyone in this part of Alaska had a lot more to worry about than caring for a dog.

After a few hours on the island, she couldn't wait to head back to Cordova. By then she'd made up her mind to leave Alaska. Madsen was sorry, but the school officials would have to find someone else to teach the remainder of the school year. She would return to Long Beach, where she'd visit with her parents and figure out the next stage of her life. Selanoff wanted to leave Alaska, too. He had fewer ties to Chenega than most villagers and at age thirty-two was ready to see more of the world. Madsen invited him to come to California. From there he could head off to points beyond.

A few days later, she found herself in Anchorage at the state education offices, where she dropped off the Chenega records. An official there tried to persuade her to teach somewhere else in the state to finish out the school year. But she had had enough. In a few days she and Selanoff boarded a plane to California.

Over those first few weeks, some of the men of Chenega made additional trips by boat back to the island. On their way they explored the shoreline along Knight Island Passage, Whale Bay, Dangerous Passage and other waters looking for signs of loved ones.

On Knight Island they found the second floor of Steve and Anna Vlasoff's house, which had been one of the few two-story houses in the village. In the violence of the tidal wave, the two floors had come apart, and the top floor had floated some miles out of the cove before washing up on the beach. Inside, the Vlasoffs' bed was still made.

On Knight Island, too, villagers found the body of Avis and Joe Kompkoff's daughter, Jo Ann. The corpse had been attacked by birds and was so damaged that Avis was not allowed to see it. But she received the cross and T-shirt that Jo Ann had been wearing when Avis heard a voice telling her to send her home to her grandmother's.

A few other bodies were found, but most of those who died at Chenega were lost forever. Those who survived had to somehow pick up the pieces, overcome their feeling of numbness and loss and rebuild their lives. The solution, they realized, was to reestablish the village. But where? There was talk among the survivors of perhaps returning to the island and rebuilding Chenega where it had been. But the idea was quickly dismissed. For one thing, no one could be certain there wouldn't be another earthquake and tidal wave someday; it seemed foolhardy to rebuild in a place that had suffered such destruction. For another, how could they live in a place where so many loved ones had died? But most important, especially to the women, was the fact that their church had been destroyed—a sure sign they shouldn't return.

The discussions about the fate of Chenega, mostly among the elders of the village, with input from Bureau of Indian Affairs officials and others, went on for weeks. One possibility, supported by the Bureau of Indian Affairs, was to move to Tatitlek, the other Alutiiq village in Prince William Sound, twenty-five miles northwest of Cordova.

Accessible only by water or air, Tatitlek was located on the eastern side of the mouth of Valdez Arm. It had a population of about one hundred. The village was on the water, but it was at a higher elevation than Chenega. Tatitlek's harbor would need to be dredged, but other than that the village had made it through the earthquake with relatively little damage.

Like the Chenegans, the people of Tatitlek lived a largely subsistence lifestyle, relying on fishing and hunting for much of their food. But the village had seen more prosperous times. In the early 1900s a rich copper deposit had been found several miles north, at a place called Ellamar. The mine provided jobs and brought more people to Tatitlek, with the population swelling to 212 in 1920. Soon after, it started to drop, gradually at first as the copper deposits were played out, and then precipitously, when a flu epidemic wiped out half the population in 1922.

A few Chenegans had relatives in Tatitlek, and the two villages shared Alutiiq heritage and culture. But most Chenegans had never had a great affinity for the other village, and vice versa. When discussing the possibility of moving there, many of the survivors said they would prefer being closer to Ellamar.

Calling a meeting of its tribal council, the Chenegans voted to move to a new village site between Tatitlek and Ellamar in late May. They would live in tents while new homes were built from materials provided by the Bureau of Indian Affairs.

Ashen, meanwhile, was working on his plan to supply the Chenegans with items that were essential to their lifestyle: boats and guns. The villagers had lost almost all their boats in the quake, and only one rifle had been found. So Ashen, after consulting with the chief, Charles Selanoff, with money provided by his Red Cross chapter back home, arranged to purchase rifles through Karl's Hardware, on Cordova's main street, for each of Chenega's fifteen remaining households. Ashen also bought hip boots and other foul-weather gear for the villagers, and ordered from August Tiedmann, the best boat builder in town, four twenty-four-foot skiffs. Ashen would also buy some sixteen- and eighteen-footers, making fifteen boats in all, each with an outboard motor. The villagers would have most of what they needed to hunt and fish.

Ashen thought that Selanoff, as chief, should get the first skiff, but Selanoff insisted that all the heads of households draw straws. The Red Cross insurance salesman had taken a liking to the Alutiiq chief. He had lost two children in the quake, yet he maintained his dignity. Ashen told a newspaper reporter later, after he'd returned home to California, that through meeting the chief he had gained new insight into people's good qualities. "Charlie Selanoff is a man I'd like to call my friend," he said.

The US Army showed up in Valdez on Saturday morning, in the form of 5 officers and 102 enlisted men from Fort Wainwright, near Fairbanks. Military commanders in Anchorage first had gotten word about the situation in the town the evening before, in the form of a short radio message.

"Valdez is a shambles" was all it said.

That had been enough to get the military mobilized. Some of the soldiers were put on helicopters that left shortly before midnight. Valdez officials, hearing of the troops' impending arrival, sent cars to the town's dark airfield so their headlights could light the landing zone. But bad weather forced the choppers down in Gulkana, 120 miles north of Valdez. They resumed the trip in the morning, landing about the time a truck convoy arrived carrying the rest of the troops and emergency supplies, including tanks of fresh water, a water-purification system and C rations. The soldiers set up a command post at Owen Meals's generating plant and commandeered the Switzerland Inn as temporary living quarters.

Both the power plant and the inn had made it through the quake in good shape. Much of the rest of the town was a mess, however, and the townspeople were unnerved. Many of them had left Friday evening, only to come back a few hours later when it seemed that things were calming down. But it had been a rough night, and now most of those who had come back were preparing to evacuate again.

The initial exodus had begun in the first hour after the quake. It was, literally, a bumpy ride, as the earthquake had left the Richardson Highway badly cracked. For some, just getting to the highway required a detour around a huge fissure on Alaska Avenue. And on the highway itself, a traffic jam developed a mile out of town where a crack had opened in front of a bridge, rendering the road impassable. Eventually the highway department came out with truckloads of sand and filled it in.

Doors were left unlocked, food was left in the fridge, valuables were left behind. The people of Valdez would deal with all of that later. Everyone had experienced personal trauma, the fear that they would be crushed by the house they were standing in or swallowed by the street they were walking or driving on. But there had been a collective trauma as well: a sickening realization, as word filtered through the streets about the destruction of the waterfront, that loved ones, friends and neighbors had disappeared, that in a horrifying instant their lives, and the life of their community, had been irrevocably changed.

For some it was all too much. The pain of the earthquake reached so deep that they left Valdez and never returned. For most, however, the departure was temporary.

Young Gary Minish and his family joined the stream of cars on the Richardson Highway that night. They all piled into the family car after their house started filling with water, but a fissure in the driveway prevented them from getting on the highway. As the yard itself began to flood, a tow truck happened by and helped them get the car on the road.

The family didn't go very far. There were rumors that another wave might hit Valdez, so they stopped at a patch of high ground at Mile 6 of the highway, along the Lowe River. It seemed as if half the town had heard the same rumors and had come to this spot, parking their cars and trucks along the road. People were in a state of panic, crying and screaming, wandering from car to car asking if the occupants had seen a husband or child, or a family pet that had run off in the chaos.

Others drove farther, through Keystone Canyon and over Thompson Pass, stopping at one of the state highway department encampments or one of the roadhouses that were a vestige of the time when travel on the highway was a multiday affair. Some made it all the way to Copper City or Glennallen, more than one hundred miles from Valdez, before stopping.

A few didn't leave at all. After the quake, Dorothy Moore's father, Sidney, dutifully reported across the street to the power plant where he worked. The building's brick facade was intact, and although the interior had flooded and had to be bailed out, only one of the three diesel generators had failed. Moore decided to stay to help get things back to as normal as they could be, all things considered. That settled it for Dorothy's mother, Edith. I'm not leaving without my husband, she said. And anyway, she went on, the house isn't in too bad shape. So the Moore family remained, and their house quickly became a gathering place for others. The first night, the eight members of the Moore family were joined by thirteen neighbors, including a boy who had lost his father at the dock.

Valdez *was* in shambles, as the radio operator had put it. There were huge piles of shattered lumber and other debris where the waves had pushed the ruins of the waterfront into the town. The shaking and rolling had done much damage to at least two-fifths of the homes and almost all of the commercial buildings; in the center of town, loose bricks and broken window glass lay scattered on the streets. Cars had fallen into crevasses, their back ends pointing to the sky. The floor of the bank had heaved up, making the prospect of opening the safe problematic. Aboveground, utility lines had been stretched and in some cases snapped, cutting power and telephone service; belowground, the fissuring had effectively shredded the town's water and sewer lines. The town's two-hundred-thousand-gallon water tank managed to remain intact despite rocking crazily, but because the water lines were cut all over town, it had quickly drained.

Fissures that had opened in the ground had continued through buildings, tearing them apart. Valdez High School, just five years old, was damaged in this way. The concrete slab that served as its foundation was rife with cracks. In the gym, where the slab was covered with a linoleum floor, a crack had narrowly missed the school's "Buccaneer" logo painted in the center. The elementary school was only slightly less damaged. Harborview, the home for the mentally disabled near the Richardson Highway, had suffered severe cracking of both its floors and roof. The fifty-some patients had been brought to the dining room, which was considered safe, and a staff member played the piano there to try to maintain calm.

When the shaking had stopped, water had compounded the damage. The first wave—the one that the *Chena* rode and that destroyed the small boat harbor—had washed inland only a few blocks beyond McKinley Street. A second wave of about the same intensity had arrived less than fifteen minutes later. Neither inundated the business district to great depth—perhaps two feet at most—but they knocked around many of the small boats, which eventually were washed out into the bay and sank. (The larger *Gypsy,* the tourist boat, may have been struck by the *Chena;* it ended up at the bottom of the bay as well.) Oil storage tanks were also damaged by the waves. Those belonging to the Union Oil Company were knocked over and began to leak, and oily water began to spread through the south side of town.

Before they left town that evening, families tried to track down loved ones. Dan Kendall, who had left Stanley Knutsen and Dennis Cunningham on the dock not long before the quake, walked over to Harborview to check on his mother, who worked there as a cook, and let her know that he and the rest of the family were all right. Gloria and Walter Day's son Pat returned after a while in the family's red pickup truck. He confessed sheepishly that he had not gone down to the waterfront to pump out their boat as promised but instead had gone for a joy ride inland. His parents were never happier to hear that he'd shirked his chores.

Some residents went down to the waterfront to look for survivors. The water was still stirred up and filled with debris, and the situation looked hopeless. One body was found—that of George Joslyn, a longshoreman who lived by himself near Dan Kendall's house. Later, when reports of missing persons were sorted out, the death toll down at the docks became clear. Thirty people had died—twenty-eight on the dock proper and two in the *Chena*'s hold. Other than the two men on the ship, Joslyn was the only one whose body was recovered.

As the evening wore on, the power plant became the center of activity. Town leaders gathered. A radio there enabled them to reach the *Chena* offshore and make arrangements for the town doctor to get to the ship. Meals also owned the local telephone service, and the exchange was in the same building. Most phones were out, but long-distance microwave and undersea cable connections had survived the shaking.

For the many residents who had parked out at Mile 6, by 10 p.m. the rumored tidal wave had not materialized. Thinking that things were calmer in town, some of them jumped in their cars and headed back down the Richardson Highway.

But the disaster wasn't over in Valdez.

Shortly before, the oil that was leaking from storage tanks caught fire. The flames spread, and soon what was left of the waterfront—including the Morgue Bar, on the remains of the earthen causeway that led to the main dock—was ablaze. Even worse, officials received word over the radio that another wave might indeed be on its way. They sent a vehicle out to drive around town with a loudspeaker to warn remaining residents to evacuate. Those leaving Mile 6 now encountered a string of cars carrying new evacuees out of Valdez, including the dozens of mentally disabled residents of the Harborview home. The drivers warned those who were heading back to town that the whole town would burn or be destroyed by a new tidal wave.

At about 11:45 p.m. a third wave did roll into Valdez, followed two hours later by another. Neither was like the mammoth wave that ravaged Chenega; they were more like fast-rising tides that reached partway into town. Valdez wouldn't be destroyed by a tidal wave, but the city would be swamped. The second wave, in particular, flooded buildings for several blocks inland with five feet or more of water.

There *was* a mammoth wave in Port Valdez, however, far out in the bay about ten miles from the town. Only a few people experienced it.

Basil Lee Ferrier, known as Red, and his son, Delbert, had motored out of Valdez on Friday in their thirty-foot fishing boat. They didn't have fishing in mind, however—they were headed to the narrows that separated Port Valdez and Valdez Arm to do some logging. There was a spit of land, called Potato Point, in the narrows about ten miles west of Valdez proper. Father and son anchored the boat two hundred yards offshore, ran their skiff to the beach and were up in the woods when the shaking began. With snow sliding all around them, they ran to the skiff to get to what they thought would be the relative safety of the fishing boat.

Just as they pulled away from the beach, though, the water dropped away underneath them, leaving the skiff high and dry. Even the fishing boat, they saw to their horror, dropped out of sight, sinking into a narrow channel as the water continued to recede. Then, as quickly as it had left, the water returned with a rush, floating both the skiff and the larger boat. Amid the turbulence, the Ferriers couldn't control the skiff, but they were lucky: it moved close to the fishing boat and they managed to climb on board.

Delbert pulled up the anchor and his father got the engine going. Just then—shortly after the shaking stopped—the boy saw a large wave developing to the northeast, in the direction of Shoup Glacier, which ended in a small bay on the northern side of Port Valdez about eight miles west of the town. Terrified, Red Ferrier turned the boat to the southwest, toward the Valdez Arm, and gunned the engine, trying to outrun the wave or at least get out of the narrows before the water reached them. He pulled out of the narrows and into wider water just as the wave, dark with mud and timbers, rushed through. Ferrier estimated its height at fifty feet or more; it overtopped a navigation light in the middle of the narrows that was on a concrete pedestal thirty-five feet high. But as it emerged from the narrows into the arm it decreased in height, enabling the Ferriers' boat to ride over the top of it—just as Howard and Sonny Ulrich had managed to do in Lituya Bay six years before.

It took several hours for the water to calm down—for a while, wave after wave came through where they had finally stopped, at a place called Jacks Bay just outside the narrows. Finally, as it grew dark, the Ferriers began heading back to Valdez. As they passed through the narrows, they saw dead snappers floating everywhere on the surface of the water.

Later, scientists discovered signs at Cliff Mine—the once-prosperous gold mine near the mouth of Shoup Bay—that water had reached a height of 220 feet above sea level. And just across Port Valdez, at Anderson Bay, there was evidence that the water had reached about 80 feet. It seemed that these large waves and the wave that the Ferriers rode out were related and perhaps were all caused by an underwater slide somewhere near Shoup Glacier.

At the time, no one was sure. But one thing that soon became clear was that the waves in the western end of Port Valdez were responsible for Valdez's thirty-first fatality. Harry Alden Henderson was a forty-nine-year-old fisherman who had a small camp on Anderson Bay. When others went to check on him after the earthquake, there was no sign of the camp, or him.

On Saturday morning, when the soldiers arrived in Valdez, they found the town leaders meeting at the power plant, trying to figure out what to do. Some of the residents out at Mile 6 had finally driven back to town in the middle of the night. They'd figured that the worst of the flooding was over and had heard that the oil-tank fire, though spectacular, probably was not going to destroy the rest of the town. So they'd reentered Valdez at about 3 a.m. Even the Harborview residents had been returned to the home.

But now, in the morning, with the damage from the high water adding to the destruction from the shaking the day before, town officials were thinking that a near-complete, and final, evacuation should be ordered. There was little or no fresh water, the sewer system was destroyed and, perhaps most important, there was a huge cleanup to be undertaken. So together with the military officers, they decided to evacuate the town once and for all. Residents had until noon to gather up their things and leave. About forty-five people, mostly men, would stay and help the soldiers with the cleanup.

Of those who left Valdez after the quake, about three hundred ended up in Fairbanks, where local residents and hotels accommodated them without charge. Others stayed in Glennallen or Copper Center; some moved in with relatives and friends in and around Anchorage. The state was insistent that school-age children finish out the school year wherever they ended up.

Some went farther afield, to the Lower 48—Outside, in the parlance of Alaskans. A few days after the quake, Gary Minish took the first plane ride of his life, along with two younger siblings, heading to South Dakota to stay with his grandparents. He became something of a celebrity in his adoptive school—he'd survived the great earthquake, and he was in such good shape from life in Alaska that he was one of the school's best athletes. Gary and his family, like other families, came back to Valdez that summer after the school year ended; other Valdez residents were away only a few weeks.

The cleanup had gotten under way in earnest on Saturday afternoon. Crews repaired utility lines, and some semblance of telephone service was restored through much of the town. Houses and commercial buildings were inspected, and some were condemned and torn down. Others were judged to be in need of repairs but otherwise safe for occupancy. The Army Corps of Engineers came in and hired contractors to clean up the debris around the waterfront.

The soldiers moved from the Switzerland Inn to larger quarters at the movie theater. A mess hall was set up, first at the inn and then at the Harborview. There was never a shortage of food, as supplies were brought down the Richardson Highway by a steady stream of trucks. Those who remained in their houses were given tanks for water and waste; these were put outside to be filled or emptied each morning by soldiers. For those whose houses had been condemned, the Salvation Army arrived with a plan for temporary housing—in the form of gleaming white house trailers—called Operation Mobile Igloo.

Throughout the weeks that followed, residents trickled back to Valdez. Life there wasn't particularly easy, and there were setbacks—including, a month after the quake, the deaths of four officers in the crash of a National Guard plane that had just dropped off the governor for a tour of the relief work. But there were signs of progress too. Among other things, Valdez once again had a newspaper, the *Earthquake Bugle.* It was little more than a mimeographed sheet, but it kept townspeople up to date on the news.

Years later, Gloria Day, who, with her husband, Walt, had come back to town just a few days after the quake, would remark on how stressed everyone in Valdez was in the weeks after the quake. No one knew it at the time, she realized, but the combination of the events of the day—the loss of so many friends and neighbors—and the rough postearthquake conditions led to short tempers. Life in Valdez after the earthquake was full of uncertainties and annoyances; to Gloria, it was a wonder they got through it.

The biggest uncertainty, of course, was what to do, in the long term, about the town itself. Henry Coulter and Ralph Migliaccio, government geologists, conducted soil tests and otherwise studied the site, but the people of Valdez didn't need scientists to tell them the obvious: if the waterfront had collapsed once, it could do so again. (In fact, Coulter and Migliaccio had found evidence that the edge of the glacial plain had collapsed several times in the previous seventy years, although not with such devastating consequences.)

The land that Valdez was built on was inherently unstable, and the town was a dangerous place to live. It would have to move. But where?

Owen Meals had a suggestion: the land that his father and

George Hazelet had homesteaded back in 1901, four miles northwest of the current town. Of the 670 acres—roughly a square mile—there was plenty of flat land that would be perfect for homes and businesses and a waterfront that could be made into a decent port with a little work. Best of all, there was plenty of bedrock that would keep the land stable in the event of another earthquake. There had been no fissuring on this land during the Good Friday quake. As further proof of its suitability, the one person who currently lived there—a crusty old man named Nicholas Mishko, who had had a falling-out with his neighbors in Valdez a few years before and in a fit of pique had hauled his home over to the homestead site—had ridden out the quake in good shape, his home undamaged.

The Meals-Hazelet land had never caught on as a town site back at the turn of the century. But now it had a certain appeal. The land was still owned by the descendants of Jack Meals and George Hazelet—there were eight heirs in all—but Owen Meals had a plan for changing that.

11

THE BARNACLE LINE

Later in his career, George Plafker proudly displayed a small plaque in his office at the Geological Survey in California. It read:

IF GEOLOGICAL AND GEOPHYSICAL DATA CLASH,
THROW THE GEOPHYSICS IN THE TRASH.

The sentiment was a bit harsh, perhaps, though Plafker didn't mean it personally. By then he had many friends and colleagues who were geophysicists, geochemists and other types of what are collectively termed geoscientists. But Plafker drew a distinction between them and him, an "ordinary" field geologist. The words of the plaque described what he felt was a truth: that he (and others like him) had an advantage over more august scientists who seldom if ever went into the field and were more comfortable working at a desk, analyzing data gathered by others and coming up with ideas that, to Plafker at least, sometimes didn't make sense.

Plafker believed that to fully understand something you had to experience it firsthand. The conviction dated from those summer field trips with A. C. Hawkins, the fill-in Brooklyn College professor, back when Plafker was first exposed to geology. Hawkins had urged his students to experience rocks in their natural settings, and Plafker had taken his advice to heart. His first job as a geologist had been disappointing, working at a desk poring over maps and reports for the military. But since he'd gone to Alaska with the Geological Survey, and even while his career took a detour to oil company work in Central and South America, Plafker had relished what being in the field gave him: a feeling that he knew what he was talking about. He'd walked the land and seen geological formations with his eyes. He knew how to read rocks, and what he learned by reading them was invaluable.

Now, in the summer after the Alaska earthquake, his work took a turn. He'd still be out in the field, seeing and reading the rocks firsthand. But he'd be learning from something else as well—a tiny sea creature.

The northern acorn barnacle, *Semibalanus balanoides,* is one of the most common barnacles in the cold oceans. Take a look at the rocks in a cove, the pilings at a pier or the hull of a boat just about anywhere in the northern Atlantic or Pacific, and chances are you'll find a lot of the little organisms. Despite the name, they don't resemble acorns so much as tiny white volcanoes, about a half-inch wide at most, with five or six hard plates making up the flanks and on top, where the crater would be, a diamond-shaped opening that serves as a kind of trapdoor. Inside are the soft tissues, including six pair of feathery appendages—some people might call them legs—that emerge when the barnacle is submerged and the trapdoor opens; they filter seawater, removing plankton and other food that is then delivered to the mouth.

Because of their hard plates, barnacles were initially thought to be mollusks, kin to clams. Then, in 1830, John Vaughan Thompson, a British Army surgeon and marine biologist, wrote of the similarities between the larval stages of barnacles and crustaceans like shrimp. Later, no less of a naturalist than Charles Darwin became involved; in the mid-1840s he became obsessed with barnacles, studying them for eight years and fastidiously describing and classifying them. There was no longer any doubt about it—barnacles were relatives of shrimp, crabs and lobsters.

Unlike those crustaceans, however, barnacles don't move around, except for the few weeks they are larvae, when they swim until they find a suitable rock or other rough surface to land on and grow. Then they use their antennas to find other nearby barnacles and settle, secreting a superglue that locks them in place for the rest of their lives—perhaps five to ten years—and beyond.

Like other barnacles, acorn barnacles on rocks or pilings live in the intertidal zone, between high and low water. For two periods almost every day they are submerged and able to feed, and when they are out of the water the trapdoor shuts to keep them from drying out. But *Semibalanus balanoides* doesn't live just anywhere in this zone; the barnacles tend to settle in a narrow band with an upper limit that is at or just a little below mean high water. That means they are submerged at most high tides (though not all—they can survive for some time out of the water), and when they are submerged it's for a relatively short time. This gives them some advantages. They are underwater long enough to get the necessary food to survive and grow, but their exposure to predators like starfish and sea snails is minimized.

When he arrived in Alaska, Plafker knew nothing about the northern acorn barnacle, or any other kind of barnacle for that matter. He was a geologist, not a marine biologist. But he was soon to become quite familiar with *Semibalanus balanoides.* The barnacles' intertidal existence—specifically, their precise location within the intertidal zone—was not just advantageous for them. It would prove extremely useful in studying the Good Friday earthquake.

After their arrival on the twenty-eighth, Plafker, Arthur Grantz and Reuben Kachadoorian stayed in Alaska for two weeks, working feverishly to collect data to compile into a report. The Geological Survey wanted to publish a preliminary study of the earthquake as soon as possible. Even if Charles Richter and other scientists differed as to the precise magnitude, the quake was already acknowledged as the strongest to hit North America and one of the most powerful ever measured anywhere. Seismologists and others around the world were interested in learning as much as possible about it. A quick report would also help build support from policy makers in Washington for a broad, and costly, program to study the quake in following years.

After a few days together in Anchorage and a few more reconnaissance flights, the three had gone their separate ways. Kachadoorian, with his interest in engineering geology, had focused mostly on the structural damage to buildings, roads, port facilities and the railroad. Grantz, who had spent more time in interior Alaska, went to the Matanuska Valley, northeast of Anchorage, between the Chugach Mountains and the Talkeetnas, to the north. Plafker said he'd take the coast—he'd head to Cordova to work Prince William Sound. He was familiar with the area from prior years and knew a lot of people there, which would be a big help in gathering information.

In fact, the previous year Plafker and a colleague had been as far as Middleton Island, the most southeasterly of the islands in that part of Alaska. Middleton was so far to the southeast—about eighty miles from Cordova—that it was outside the sound and in the Gulf of Alaska proper. Plafker was anxious to see it again.

He was anxious, actually, to see as much of the area as he could. The reconnaissance flights out of Anchorage suggested that the land all over southern Alaska had deformed during the quake, rising up or sinking down. The changes were quite dramatic, as the three men had seen up close on April 1, when they had flown to Homer. The town was past Seward at the tip of the Kenai Peninsula and included a spit of land that jutted into Kachemak Bay. When they arrived, a work crew was busily jacking up an old inn on the spit. Before the quake, it had been comfortably above the high-water line. Plafker looked at the cribbing of large timbers that the men were putting under the building as they raised it. It was about six feet high; the land had sunk that much.

Working out of Cordova, even for just a few days, he'd be able to see more of this kind of deformation up close. And he'd get a good idea of the work that would be needed when he and others came back in a month or so to make a detailed survey. Just as important, he'd get survivors' accounts when they were fresh and most accurate—before they'd shared stories with others and inadvertently come to mix other people's perceptions with their own.

Plafker enlisted Jim Osborne—the Cordova Airlines pilot who had rescued the survivors at Chenega—to take him around the sound. He met up with Osborne on the thirty-first, when he and Grantz stopped in Cordova during another of their reconnaissance runs. Plafker found the pilot to be a very useful source of information about the effects of the quake. After all, Osborne flew the sound twice a week to deliver the mail at almost every village and outpost, so he knew the land and the people as well as, and in most cases better than, anyone. He told Plafker about what he'd seen and heard during those first couple of days, including reports of tidal waves that had not made any news accounts, at out-of-the-way places like Peak Island and Port Oceanic. And he agreed to let Plafker accompany him on his mail run later in the week.

So on April 4, Osborne and Plafker took off in the Cordova Airlines Widgeon on what amounted to a grand tour of the sound. While Osborne dropped off the mail, Plafker talked to survivors about the quake and jotted down information about the land deformation in his field notebook. In most cases he didn't have time to do any actual surveying, but he didn't need to. People who live by, and off of, the sea know their tides, and in this part of Alaska the tides had been altered by the quake. They told Plafker of the changes, and that was all Plafker needed to know, for now, about what had happened to the land.

At Tatitlek, the first stop, the village postmaster told him that the high tide was four to six feet lower than it had been before the quake, indicating that the land had risen that much. Other than that, though, Tatitlek had not been badly affected. The shaking had caused little damage—though it had made the church bell ring—and people had not been knocked off their feet, as they had elsewhere. The water had receded from the shore at one point, but unlike at Chenega, there had been no tidal wave.

The story was much the same at the next stop, nearby Ellamar. There had been no tidal wave and no one had been hurt, although there had been some damage—a chimney had collapsed and five windows were shattered. A man named Carl Aranson told Plafker that the quake was the worst he'd ever felt, that the shaking had made the furniture move around inside his house and had caused him to fall over when he ran outside.

In all, the Widgeon made seventeen stops that day. Plafker interviewed people at almost every location and took photographs everywhere. He and Osborne visited Port Nellie Juan, where the three members of the Chimovisky family had died, and other canneries at places like Port Oceanic and Port Ashton, where the winter watchmen had survived and told Plafker what they'd experienced. The Widgeon landed at Peak Island, practically in the geographic center of the sound, and at the two big islands along its southeastern edge: Hinchinbrook and Montague. When no people were around to tell him how the tides had changed, Plafker occasionally surveyed the shoreline himself. Osborne served as his rodman, or assistant, standing near the water's edge with a stadia rod, essentially a long ruler. Plafker, on higher ground where he found evidence of the prequake tide line, sighted through a hand level—a small scope with a leveling bubble inside it—and read the markings on the rod, subtracting six feet for his height. Often he wouldn't even need to use the leveling bubble—he could just line up the scope with the far-off horizon over the water and be assured that he was getting a level measurement.

In the afternoon they flew back to Cordova to refuel for a longer flight southeast into the Gulf of Alaska, to Kayak Island—where the Coast Guard lighthouse keeper had lost his life—and, finally, Middleton.

Middleton Island had long been home to small groups of fox

farmers and, for about five years until it was deactivated in 1963, an air force radar installation. A mere four miles long and a mile wide, Middleton sits alone in the middle of the gulf, directly exposed to the Pacific, which comes rolling in on the Japanese Current. As a result, plenty of driftwood and other debris washes up on the island. Plafker remembered this from the year before—piles of graying driftwood at the back of the island's beaches, where they had been left by high winter tides.

When Osborne eased the Widgeon down at Middleton, one look at the shoreline told Plafker all he needed to know about the earthquake's impact. Those driftwood piles he'd remembered from the summer before were now much higher; they looked to be about eight feet above the high-tide line. The waves would never reach them again.

Plafker returned to Anchorage on April 6 and, after a final reconnaissance flight to Kodiak, spent his last few days interviewing survivors, including Kris Madsen, who was soon to leave Alaska for good. She told him what she had seen of the tidal wave at Chenega from her vantage point above the schoolhouse. Plafker also interviewed some of the surviving villagers, and witnesses to tidal waves elsewhere, and was developing an understanding of what had happened at the various communities around southcentral Alaska. But a lot was still unknown about why events had unfolded as they did.

On April 10 the three geologists packed up their notebooks, maps and photographs and boarded a Pacific Northern Airlines flight for San Francisco. They had been on the ground in Alaska for thirteen days. Now their job was to write about what they had learned about the Good Friday earthquake.

At the Geological Survey's offices in Menlo Park, they kept interviewing survivors and witnesses by phone. Plafker talked with Captain Stewart of the *Chena* and an official in Crescent City,

California, among others. But mostly he and his colleagues compiled data, gave it to a team of cartographers and chart makers assembled for the project and started writing. Grantz handled much of it, but all three of them participated, working essentially nonstop for two more weeks. On April 27, just a month after the quake, the Survey published Circular 491, "A Preliminary Geologic Evaluation" of what it called "Alaska's Good Friday Earthquake."

The report ran to thirty-five pages, twenty of which were maps and photographs. On the title page, below their three names, were the place and date of publication: Washington, 1964. Plafker, who was proud of what they had accomplished, thought it should have been more specific: April 27, 1964, to show readers just how fast they'd done the work.

In plain, straightforward language, the opening paragraph set forth what had happened: "At 5:36 p.m. on Good Friday, March 27, 1964, a great earthquake with a Richter magnitude of 8.4 to 8.6 crippled south-central Alaska. It released twice as much energy as the 1906 earthquake that wracked San Francisco, and was felt on land over an area of almost half a million square miles."

The circular detailed the immediate effects of the quake—the rockslides and avalanches, the cracks and mud spouts and the cracked ice and pressure ridges on frozen lakes. It inventoried the damage to the state's roads and railroads, dock facilities and canneries and schools and other public buildings. (Damage to military infrastructure and, certainly, damage to the nuclear missiles in Anchorage were not part of the report. The Pentagon would deal with all of that on its own.)

Circular 491 described the destruction in Anchorage, Seward, Whittier, Valdez and other communities and noted the cause of it: seismic shaking, slides, tidal waves or some combination of all three. The geologists had been able to learn more about what had happened in some communities than others. They had a good understanding of the slides in Anchorage, for example: they included a fairly detailed description of the role of the slick Bootlegger Cove Clay and even a three-step, cross-sectional diagram that showed how the Turnagain Heights slide had progressed. They knew that the waterfronts of Valdez and Seward had collapsed because they were built on unconsolidated glacial sediments. They realized that Kodiak had been hit by "seismic sea waves," as they referred to them, which had also traveled across the Pacific and to the California coast. But among other uncertainties, they were still unsure about the source of Chenega's destruction. The village, they wrote, "was hit by a large wave of unknown origin."

The geologists also put to rest concerns that they themselves had had when they first came to Anchorage: namely, that some rivers might have been blocked by landslides, which could have led to catastrophic flooding. They had found no evidence of this in their reconnaissance flights and fieldwork. They *had* discovered an odd effect of the quake on several smaller rivers, including Ship Creek in Anchorage—water flows had been sharply reduced for about a week. Perhaps fissures had formed in the riverbanks that allowed much of the water to seep into the ground, the geologists wrote, or all the shaking had made the riverbeds more porous, with the same result. All in all, though, Alaska's rivers had made it through the earthquake in good shape.

The report also suggested a likely explanation for the phenomenon that mystified all who saw it: the dead red snappers that floated by the thousands on the water in some parts of Prince William Sound shortly after the quake. Like other bottom-feeding fish, snappers live in deep water—which in the case of the fjords and channels of the sound can be four hundred feet or deeper. They have an air bladder, filled with gas that dissolves out of their blood, that keeps them buoyant at depth, where the water pressure is high. But if the fish rise to the surface too quickly, the rapid decrease in pressure causes the gas in the bladders to expand very rapidly, killing them. The geologists proposed that currents created by the shaking and sliding during the earthquake stirred up the bottom waters, forcing the snappers toward the surface and their doom.

Plafker, Grantz and Kachadoorian devoted much of the report

to the biggest impact of the earthquake: the rising and sinking of the land over a wide swath of south-central Alaska. Although their survey efforts had been of necessity incomplete, they calculated that at least thirty-four thousand square miles of land were affected, an area the size of Indiana, and divided the territory into three zones. One was an area of "tectonic subsidence," which consisted of Anchorage and environs and almost all of the Kenai Peninsula—including Portage, which they'd seen from the plane on their first flight—and Kodiak Island. The land in this zone had subsided, or sunk, by more than five feet in some places. A second zone, of "tectonic uplift," covered the southeastern half of Prince William Sound and, on the mainland, the Copper River delta southeast to Kayak Island. Here the land had risen by as much as seven and a half feet—and when more detailed surveying was undertaken in the summer, there was little doubt that areas of even greater uplift would be discovered. The third zone was described as a "tectonic hinge" between the zones of uplift and subsidence, a roughly twenty-five-mile-wide strip running southwest to northeast that included much of the northwestern part of the sound. But what was the nature of this "hinge"? Was it a flexible one, in which the strip of crust acted like a piece of rubber holding the two bigger zones together? Or was it inflexible, and the location of a sharp break between those two bigger zones, in the form of one or more faults?

The geologists were inclined to believe that the hinge was a "zone of flexure," as they put it. For one thing, Plafker had found a location in the hinge zone where there had been no change in land level during the quake—on Perry Island, which he had visited with Osborne. If there had been a sharp break between subsidence and uplift, one wouldn't expect to find *any* areas that were unchanged.

But of even greater significance was that the three men had found no sign of faulting at the surface. The lack of any evidence of a fault had intrigued Plafker during that first reconnaissance flight on the Sunday after the quake, and even after more flights over the following ten days he and the others hadn't seen anything in the hinge zone—no obvious scarps or linear traces in the earth or snow—to suggest a fault. Of course, they had only been there for two weeks, and they might have missed it. To cover themselves, in the report they noted that "a surface break could easily have been overlooked, especially as new snow had fallen."

Plafker remained intrigued by the absence of a visible fault. Clearly there was still much to be learned about what this earthquake did to the land. And just as clearly, he was beginning to realize that there was much to be discovered about *how* the quake did what it did. He was eager to get back to Alaska and learn more.

For all of George Plafker's skills in the field—his ability to work in rugged backcountry Alaska, navigating rivers and hiking glaciers, setting up spike camps, living off C rations and fending off the occasional bear—the summer of 1964 was a very different experience. It was the summer he spent on a ship.

Not a ship, precisely—a flat-bottomed motor barge, of the kind that prowls the Alaskan coastline, delivering goods to isolated shoreline communities. This one was about eighty feet long, with a wheelhouse and bunkhouse on the deck and a mast and boom for loading supplies. The crew consisted of a captain and a seaman who doubled as a cook, and there were berths for four researchers and assistants. It was owned by the Geological Survey—they had picked it up on the cheap after its goods-hauling days were over—and had spent most of its summers taking scientists around the fjords and bays of southeastern Alaska. It had been laid up in Seattle for the winter, and was rushed back up to Alaska for earthquake research.

Its name was a familiar one to Plafker: the *Don J. Miller.*

Starting in the early days after the quake, the Geological Survey had been working to set up an ambitious research program for the summer, with scientists recruited from throughout the agency. George Gates, a former head of the Alaska branch, had been chosen to coordinate the field program. Two geologists were assigned to study the Anchorage slides in detail; two more worked in Seward and Valdez; and two worked with the Alaska Railroad. But there were many others, among them engineering geologists, soil geologists, marine geologists, hydrologists, experts at making gravity measurements and researchers who had made careers of studying landslides. What was lacking were many scientists who had studied earthquakes—it was still a fledgling field of research at the Survey.

The plan was to produce a series of in-depth professional papers on all aspects of the quake that would document and analyze what had occurred and, more important, help scientists and engineers prepare for future ones, in Alaska and elsewhere. Eventually more than fifty researchers would be involved in the publication of the various papers. There might be some doubt as to whether the Good Friday earthquake was the most powerful of all time, but there was no doubt that it would be the most studied of all time.

Grantz had wanted to stay involved, but his wife had health problems that forced him to remain in California that summer. Kachadoorian did return to Alaska to focus on the damage to and reconstruction of the state's roads. And Plafker had signed up to continue studying the uplift and subsidence, focusing on Prince William Sound and part of the Gulf of Alaska. So on May 18, after a little more than a month back home with Ruth and the children—and having spent much of that time at the Survey offices in Menlo Park—Plafker boarded a flight to Seattle. The next day it was on to Juneau and then Cordova. After a couple of days of interviews and reconnaissance—including another flight in the Widgeon with Jim Osborne—Plafker met up with the crew of the *Don J. Miller* and his fellow researchers and headed out into the sound. Their destination that first day was Knight Island, near Chenega.

Plafker had come to understand that for the purposes of his research he had been fortunate in where this earthquake had occurred. Here was a huge area of land that had been deformed by the quake, and it was relatively easy to study. Some of it was mountainous terrain that was difficult to reach, but much of it was on and around large bodies of water with plenty of shoreline where changes in elevation could be measured. With its many bays, fjords, channels and islands, Prince William Sound alone had close to four thousand miles of coastline. The *Don J. Miller*, Plafker realized, would make most of it easy to reach. And measuring the changes in elevation along it would be made easy by something else: the barnacle line.

Plafker had first learned of the barnacle line during his two weeks in Alaska immediately following the quake and had talked to marine biologists then to better understand how barnacles fit into the environment of the Alaskan coast. The idea of using barnacles to measure uplift and subsidence was not new—it had been employed in a previous Alaska earthquake, near Yakutat Bay, southeast of Cordova, in 1912—but it's fair to say Plafker and his colleagues perfected the technique and used it to a far greater extent than previously. They made hundreds of elevation measurements, the bulk of them using the barnacle line.

The concept was relatively simple. Because northern acorn barnacles establish themselves at a certain spot on rocks and pilings—at or close to mean high water—they could be used as a reference point to measure both uplift and subsidence. In an area where the land had risen up, the prequake barnacle line would now be higher than it was before, and out of the water. After a few weeks the barnacles would have died, but their white-colored plates remained, firmly cemented to the rock or wood. For areas where the land had sunk, the barnacle line would now be underwater most or all of the time. Either way, to determine the amount of elevation change, in most cases all that was needed was to know the stage of the tide—which the US Coast and Geodetic Survey had been busy recalculating all over Alaska after the earthquake—and then measure from the waterline to the top of the barnacle line.

Say the land had risen up. And say the measurement was being done at a time of day when the tide was two feet below mean high water. The amount of uplift would be equal to the measurement from the waterline to the barnacle line, minus two feet. A similar measurement could be made to determine subsidence, although in that case the measurement would be made from the waterline *into* the water to the barnacle line.

In practice things could be a little more complicated. The Geodetic Survey's tide heights and times were precisely determined at relatively few stations, so tides elsewhere had to be interpolated. And depending on local conditions, barnacles might establish themselves a little lower relative to mean high water than elsewhere. But Plafker learned to adjust and adapt as necessary, and even if the measurements had a margin of error it was small enough that the overall trends in uplift and subsidence were still obvious.

Later in the summer the work became easier, and Plafker found that often he didn't need to worry about the tides at all. Late summer was when juvenile barnacles, which had hatched after the quake and developed, settled down for good—at the new, post-quake mean high-water line. Then Plafker would have two barnacle lines—before and after—and determining the elevation change was simply a matter of measuring the distance between them.

But in some cases—especially where coastal rocks were regularly exposed to strong waves—there would be no barnacles at all, old or new. Plafker found that in those locations he could use another marine organism in their place—a type of seaweed, *Fucus distichus,* or rockweed. This grew in an olive-green band on the rocks, and the upper edge of the band correlated well with the barnacle line. When it died, it turned brown.

At some spots along the shoreline there were bands of rockweed and barnacles and, above the barnacles, a third band of a dark-gray lichen that encrusted the rocks. So for Plafker and the other researchers, hunting for the barnacle line sometimes meant looking for rocks that had a tricolor band of gray, white and brown.

The bands were usually easy to see from the *Don J. Miller.* The barge would cruise the coastline, and the researchers would look for good spots to take measurements. After a while they became experts at sighting good locations from far off. Rock formations that were smooth and closer to vertical were best, because the barnacles were close together and the band was more distinct. On gentler sloping rocks the barnacles were scattered over a larger area and the line was not as clear. Tidal pools and other features of the shoreline could also confuse things.

When they found a good spot, Plafker and an assistant—often Larry Mayo, a young geologist who had just started at the Survey the year before and would go on to have a career studying glaciers—would get into one of two skiffs and motor over to the shore to take measurements and photographs. Jim Case, a Survey biologist, would climb into the other with another assistant and work a different part of the shoreline. They tried to get accurate measurements every one to five miles, which would give them enough data to create a contour map of the region showing the uplift and subsidence.

After so much time in backwoods Alaska, Plafker was quite happy to spend a summer on the water. It was a far cry from his usual fieldwork—they weren't doing any real geology, where they'd have to stop and examine rocks and break off samples with a rock hammer and haul the samples back to their camp. But in many ways it was easier. Whether out of guilt or just habit, at times Plafker would find himself studying the geology of a location anyway. He couldn't help himself.

The accommodations were more comfortable than a spike camp. The barge was no luxury yacht—Plafker would refer to it as "that crummy barge"—but it was a pleasant change to spend one's nights surrounded by four solid walls rather than a canvas tent. With its flat bottom, the barge rolled a bit in rough seas, but in more placid water or when the captain, John Stacey, anchored for the night near shore, it could be almost relaxingly calm.

Then there was the food. The other member of the crew, John Muttart, turned out to be quite a cook, and Prince William Sound produced a fabulous bounty of seafood. When Plafker and the others left in their skiffs in the morning, Muttart would throw traps over the side of the barge, and when the men returned for lunch or dinner there would be shrimp or Dungeness crab for everyone. There was a charcoal grill on the deck if the men preferred their crustaceans barbecued. Occasionally, just to change things up a bit, Muttart would let out a line and within minutes pull up fish off the bottom for the night's meal.

They couldn't live by seafood alone—although it was tempting, Plafker thought, especially given all the Dungeness crab—so every once in a while they would radio to Cordova and place an order for groceries, which would be delivered by floatplane, piloted by Osborne or someone else. And once every few weeks they'd have to put into Cordova or another port for fuel.

As field research went, it was not a bad situation. But once or twice things threatened to get out of hand. Plafker was listening one day when one of the other researchers, on the radio to Cordova, calmly added a case of whiskey to the grocery order. Plafker blew his top, reminding the man that as Geological Survey staffers they weren't supposed to have *any* alcohol, much less a case of whiskey, on board. And the radio was on an open channel, meaning just about everyone else working in the sound, including fishermen, Coast Guardsmen and other Survey researchers, could hear the order. It was canceled.

The barge worked the sound and Resurrection Bay in the Gulf of Alaska most of the summer. For Plafker and the others it was a very productive time. They visited almost every island, including Chenega, Evans and Knight, and went up most of the fjords and channels. They also had occasional use of a helicopter to take them farther afield.

They were still interviewing survivors where they could and

studying the damage caused by waves at Chenega, Whittier and other places around the sound. Plafker and Mayo came up with a "wave magnitude scale" to quantify the water's destructive force. Magnitude 1 were small waves that ran up just a few feet above extreme high water, could float houses and could break small limbs on trees. Magnitude 3 waves could run up to 55 feet above high water, break trees up to eight inches in diameter, move heavy equipment around and deposit cobbles on hillsides. At the top of the scale was magnitude 5—waves that ran up to 170 feet and destroyed everything in their path, scoured hillsides bare and threw large boulders about like marbles.

But the elevation work—the barnacle line—consumed most of their time. Plafker and his colleagues ended up making more than eight hundred measurements around the sound. Elevation changes ranged from subsidence of almost seven feet near Whittier to a maximum uplift, on Montague Island, of nearly thirty-eight feet.

It was on Montague, the long island that marks the southeastern limit of the sound, that Plafker finally found faults related to the earthquake. Jim Osborne, in fact, had first clued him in—he had noticed something amiss on Montague from the air. When Plafker and the others surveyed the island's northwest coast, they found a long fault, running more than twenty-two miles, that had created a fresh scarp, a cliff that was more than twenty feet high at its maximum. Another fault, nearby, was shorter, and its scarp was not as high.

Plafker didn't quite know what to make of these faults. It seemed clear that they had made the uplift at Montague Island much greater than elsewhere. But it was also obvious that they were too small, and too far to the southeast, to have been responsible for the earthquake. It was more likely, he figured, that the movement along these faults accompanied the major fault break, which had occurred elsewhere.

It was all food for thought. At the end of that summer, as other researchers prepared to go back to what they had been doing before the quake, Plafker found himself dwelling on it endlessly. Seeing the earthquake's immense power firsthand—its ability to, quite literally, move mountains—had made him eager to know more. Rather than spending the fall and winter planning fieldwork for his regular job, he would spend the time preparing a report on the tectonics of the earthquake—how the earth had moved. It was an immense job, and he was, as he was fond of putting it, just a mineral resource geologist with no earthquake expertise. But someone had to do it, and the others were all caught up in their own little worlds. So it was left to him to assemble the data he and the others had collected and make sense of it. Then he had to set it all down in writing and, as best he could, come up with the answer as to why the earth had done what it had done.

It was a huge challenge, but he already had some ideas.

12

REBUILDING

The envelope was addressed, rather incompletely, to "Governor of Alaska, Alaska," but it managed to arrive in Governor Bill Egan's office in Juneau just the same. When it was opened, it was found to contain a dollar bill along with a note, scrawled in blue ink on a sheet of lined paper, from an unidentified man in Arlington, Washington. The note read in part:

> *Dear Mr. Governor,*
>
> *I am sending 1 buck to help the good people of Alaska. I am 70 years old and I think if all the people in the good old United States would just give a buck things in Alaska would come along in nice shape. . . . I'm glad I am an American and not a communist and never will be.*
>
> *God bless you all.*
> *An old Great Lakes sailor*

The one-dollar donation was just one of thousands for emergency relief in Alaska. Letters poured into Egan's office containing cash or checks, most for ten dollars or less, from all over the country—from the pupils and teachers of Edgewood Elementary School in Scarsdale, New York; members of the Goldwater for President Committee in Washington, D.C.; the Junior Women's Club of Burbank, California; and many others, including hundreds of individuals. Money came in from overseas as well, especially from Japan, which was no stranger to disastrous earthquakes.

With their donations, some children expressed concern about the fate of Santa Claus (they received a reply assuring them that he was unaffected by the quake). A seven-year-old boy in Port Chester, New York, wanted to donate his Easter basket to the cause, but when his father told him that the eggs would spoil he gave a dollar instead. Some people went to great lengths to help, among them Cyra G. Renwick of Ligonier, Pennsylvania, an amateur composer, who sent in fifty-seven dollars she had raised going door-to-door selling sheet music for a song she'd written, "Hymn to Alaska."

In all, the governor's office received more than $125,000 in private donations to help with relief and rebuilding. It wasn't nearly enough, of course. Alaskan officials realized almost as soon as the shaking stopped that no state, much less their young and undeveloped one, would have the financial resources to cope with a catastrophe of this size. Egan had made an early off-the-cuff estimate that rebuilding would cost $500 million, and he and others were on the phone to Washington immediately to ask for federal help of that magnitude. President Lyndon B. Johnson responded on the day after the quake by declaring the state a disaster area, freeing up emergency money, and sending a personal representative to the state, Edward A. McDermott, who was head of the Office of Emergency Planning. Within a week Egan was on a plane heading for Washington to discuss longer-term aid face-to-face with the president, and within two months Congress had taken up and passed an Alaska Omnibus Bill designed to free up much more money. By then estimates of the disaster's cost had declined to a little bit more than $310 million, not counting the loss of personal property. In all, the federal government ended up providing more than that amount in aid. Given the scope of the disaster—unprecedented for peacetime in the United States—Johnson also established a reconstruction commission, made up of cabinet secretaries and agency directors, including McDermott, to oversee the federal government's efforts.

Alaska needed rebuilding help in a hurry, as the summer construction season was approaching and it didn't last long. It was extremely difficult, and far costlier, to try to build in the state during the cold months amid dwindling daylight. The Army Corps of Engineers was at work almost immediately, clearing landslides that blocked the Seward Highway and other roads, installing temporary bridges where needed, cleaning up the damage in hard-hit ports and demolishing condemned buildings in Anchorage. Before the end of summer, the corps, contracting out much of the work, managed to restore some semblance of normalcy to the ports of Cordova, Kodiak, Whittier and Seward—building temporary dock facilities where needed and, in Cordova, dredging the uplifted harbor to its original depth. In Valdez, the corps quickly rebuilt a dock so that the ferry *Chilkat* could resume service. In all, the corps spent more than $110 million on work related to the earthquake.

Mostly on its own, the Alaska Railroad undertook the task of repairing its badly mangled line. A less-damaged stretch between Anchorage and Palmer, to the north, was reopened in about two weeks, while it took more than a month to make temporary repairs to the rails south of Anchorage. By summer the line was open all the way from Seward to Fairbanks.

In Anchorage, recovery hinged to some degree on the status of the areas where slides had occurred. At these locations—including Turnagain Heights, Government Hill and Fourth Avenue—the Corps of Engineers with characteristic efficiency quickly removed damaged and demolished structures. But it was yet to be determined if the areas could, or would, be made safe enough to build on once more. The acreage involved was a not-insignificant portion of the city and made up parts of some key neighborhoods—Fourth Avenue, for one, was in the heart of downtown—and if the areas could not be made safe, Anchorage's growth might be severely hampered.

Initial reports, made within a month of the earthquake by a volunteer group of geologists, were not promising. The Anchorage Engineering Geology Evaluation Group, as it was known, essentially ruled most of the slide areas off-limits. Needless to say, this didn't sit well with Anchorage's political and business leaders, or with the many residents and small business owners whose lives and businesses were in limbo.

A second group—known as Task Force 9, as it was the ninth of a series of committees set up by the federal reconstruction commission—took a more flexible approach. The task force's job was to advise where federal money should, or should not, be spent to stabilize soils and whether it made sense to repair or completely rebuild facilities on a case-by-case basis. In Anchorage, the task force first divided the city into areas of high risk and, essentially, no risk. This was similar to what the earlier group had done, but then the task force looked more closely at the high-risk areas and determined that, actually, some of the land within them could be put in a lower-risk category—and could be built on again—if it was stabilized. On Fourth Avenue, the method of stabilization that was eventually proposed was a complex and costly one, involving a wide gravel buttress running for eight blocks. It would cost $10 million.

By the summer of 1965, the buttress was yet to be built. The north side of Fourth Avenue, which had dropped ten feet during the quake, was now just a series of vacant pits. Little had been done in the other slide areas as well.

But in some ways it didn't matter. Task Force 9's recommendations were just that—recommendations, not requirements—and some Anchorage developers disregarded the risk and built where they pleased. Among them was Wally Hickel, the future governor, who built a $4 million, nine-story hotel just off Fourth Avenue near the slide area. Never mind what the task force advised, he said—he had his own engineers, and they had assured him that the ground was safe.

In fact, by a year after the earthquake Anchorage was undergoing a building boom, fueled by the desire of almost everyone to remain and rebuild the city. Building permits for projects worth $25 million were issued in 1964, three times as much as in 1963, which had been considered something of a boom year itself. Other hotels were in the works, and new government buildings were under construction. Apartment complexes were rising around the city, and a large shopping center was being built out near the airport. Perhaps most symbolic was the announcement by J. C. Penney that it planned to rebuild its store in the same location, with the same amount of retail space as before. It would be a little shorter, though—three stories instead of five.

There had been talk after the earthquake that the disaster afforded Anchorage a singular opportunity to change the course of its development. Why not take what nature had done and go with it, and build a different city out of the rubble? Planners latched on to the idea and among other things proposed a rethinking of downtown—moving it and creating pedestrian malls and other features to drastically change the city's look and feel.

By the summer of 1965, it was clear that the people of Anchorage, and their business and political leaders, would have none of it. That July, a newspaper reporter asked one local planner, E. Jack Schoop, what had happened to all those plans. "Most of it went pffft," Schoop said. Outside of the slide areas, which were still a work in progress, most of the planning, he said, had "pretty well been hammered out by property owners and the City Council."

Anchorage had been built in a largely helter-skelter fashion, and from the looks of things, despite the very real horrors of the 1964 earthquake, it would continue to be built that way. That was the Alaskan way, after all, encapsulated in Hickel's line about not letting nature run wild. After the earthquake was the same as before it: Alaskans were determined to carve their civilization, on their own terms, out of a vast untamed landscape. The earthquake was just a different manifestation of Alaska's wildness, and Alaskans would be damned if they would let it get the better of them.

In late May, after two months of living in the basement of the Baptist church in Cordova, a few of the men of Chenega journeyed to

Tatitlek to help set up a tent city and unload material for seventeen new homes. Things didn't go exactly as planned.

In their negotiations with the Tatitlek leaders and officials of the Bureau of Indian Affairs, the Chenegans had made clear—or at least thought they had made clear—that rather than settle right next to Tatitlek, they'd prefer to keep some distance, to help maintain their own village identity. Perhaps midway between the village and the abandoned mine at Ellamar, three miles away, would be good.

It's not clear what went wrong. One account suggested that perhaps the details of the agreement had never been completely firmed up, or that the word *midway* had been somehow unclear. Whatever the reason, when the advance crew from Chenega was shown the sites for their tent city and new homes, the plots were adjacent to the existing village. When the rest of the village showed up a few days later, some aboard the *Marpet* or the *Shamrock* and others by plane, no one objected publicly to the location. But some said later that when they saw the home sites plotted out with stakes in the ground it was the first time that they were certain where they were going to live.

The arrival of the earthquake refugees marked the beginning of a partnership between the two villages—which at least in legal circles would be known as Chenega-Tatitlek. But it was an alliance that over the long term, despite good intentions, did not work out.

The people of Chenega did voice their unhappiness about one element of their new homes—the size of the plots, which they felt were too small at fifty feet by fifty feet. But the homes themselves would be built of plywood and would be much more solid and up to date than what had been destroyed in Chenega. They could live with the small plots, as they would once again be able to hunt and fish, which they had largely been unable to do while in Cordova. Now, even before their homes were built, they had rifles to replace those lost, and the skiffs that Pete Ashen had ordered. The villagers could get out into the woods and onto the water.

That first summer they lived in green army-surplus tents and, with the fish camp at Port Nellie Juan wiped out, explored nearby fishing grounds. The catch was good, and they were able to sell their haul to a relatively new, and apparently insatiable, customer—the Japanese. They made about as much money as they would have had the Nellie Juan cannery been in operation.

By summer's end, Chenega's schoolchildren were enrolled in Tatitlek's school. A team of fifteen carpenters hired by the Bureau of Indian Affairs showed up in September and with the help of the villagers began building the new homes. The federal public health service also came in to design and build a public water and sewer system, something that Chenega had never had. In October, the Tatitlek harbor was dredged to pre-earthquake depths, making life easier for those with larger boats like the *Marpet* and the *Shamrock*.

Outwardly, at least, all appeared to be going well in the combined community. In November the two village councils joined together for an all-community meeting to plan a new, larger church that could handle a population that now numbered about two hundred. There was also talk of building an airfield and a breakwater.

But all was not right among the Chenegans. As Nick Kompkoff—who eventually became a Russian Orthodox priest—put it a decade later, while the decision to move had to be made, and at the time Tatitlek was the best (if not the only) option, for the villagers it was a decision that came too soon. Emotionally they weren't ready; the wounds were too fresh.

For some, the hurt showed itself in complaints about little things. Avis Kompkoff hated the time they spent living in a tent and the camp-style breakfasts; for a while, it seemed, she had to eat powdered eggs every day. But others had deeper concerns. Mary Kompkoff, Nick's wife, thought that the fundamental problem was that the new village was not theirs. Occasionally, she recalled later, Tatitlek residents would make this clear to them, saying: "You can't do this. This is not your village." But the distrust and dislike came from the Chenegans too. "We couldn't—I don't mean to say we didn't—get along," she said. "We weren't used to living with Tatitlek people."

A few of the villagers had decided not to make the move to Tatitlek, opting instead to settle near family in Cordova or Anchorage or elsewhere. But some of those who did make the move began to drift away that fall, even as plans were made in their new community for the larger church and other improvements. Nick Kompkoff himself went to Anchorage to study for the priesthood, though he eventually returned and led the Tatitlek church for three years. Others went to Cordova and, eventually, Valdez. At least one Chenegan moved Outside, to Seattle.

Paul Kompkoff Sr., who had moved from Chenega to Cordova before the quake, recalled later that Tatitlek "wasn't right for some of the people. . . . They were away from the hunting and fishing that they knew." And Nick Kompkoff would say that on some level, through no fault of their hosts, the people of Chenega had not felt at home. "We still wanted our own place."

The diaspora had begun. Eventually there would be more Chenegans in Anchorage than in Tatitlek or anywhere else. Chenega would become a community in exile—tied together by background and culture and the nightmarish events of March 27, 1964, but without a physical location of their own.

It would take two decades, and a lot of starts and stops, for the community to find a home again.

The announcement appeared in the June 20 edition of the *Earthquake Bugle,* the mimeographed sheet that a local couple, Ed and Frances Walker, put together weekly to keep Valdez residents informed about recovery efforts. In a show of solidarity with the women of Valdez, the announcement said, the Lovable Company, a lingerie manufacturer based in Los Angeles, had sent twenty-five dozen brassieres to the town along with various foundation garments and garter belts. The gift had made quite a splash nationwide—it had been written up in newspapers from Connecticut to Arizona—but the shipment had been lost for a while, stuck in a warehouse in

Fairbanks. Now the bras had finally been delivered and were available for the taking at the Hotel Valdez. The *Bugle* thanked Lovable for "this useful yet highly flattering gift" and exhorted its female readers: "Line up, gals!"

There wasn't exactly a stampede to pick up free undergarments. For one thing, there weren't yet three hundred women in Valdez. For another, the women who *were* in town—and the men too, for that matter—had a lot of other things on their minds and a lot of work to do.

The city was going to be moved, that much was certain, but what was unclear was when, and what it would mean for each household. Until the relocation happened, those now in Valdez—the original forty-five who stayed had been augmented by several hundred more, and more people were returning every day—had to make sure their homes could make it through the winter. Or if not, they had to obtain one of the trailers being offered.

A month after the quake, the Valdez town council voted to move. But moving an entire town had almost never been done in the United States, and such an undertaking was far beyond the capabilities of local leaders. The state housing authority was put in charge, with much assistance, and money, from the federal urban renewal program. The Army Corps of Engineers would do most of the basic work of building the new town, including extending the Richardson Highway four miles to the new location.

Owen Meals had proposed donating 110 acres of the 670 acres of the old Meals-Hazelet homestead, four miles to the north. Together with other land that Valdez could annex, there would be more than enough room for the population of old Valdez and then some. Meals didn't want any payment, just a waiver of taxes on the land that remained. There was some grumbling about this idea among the people of Valdez—might Meals, who owned the power plant and telephone service in the old town, have something up his sleeve? But eventually it was accepted, and a planner was hired to come up with a design for a new Valdez. The planner, thirty-three-year-old

Paul Finfer, came from a place about as far removed from Valdez, physically and geographically, as possible: Mishawaka, Indiana, near South Bend.

Finfer didn't have much time; officials had decided that they should have a plan in place by the end of summer, to get started on what would be a lengthy relocation process. He got to work, at first by sending out a questionnaire to determine residents' desires for a new town. They were asked what kind of lot size and shape they would prefer, whether they were planning to build their own house, whether there should be standards for landscaping and maintenance, and other basic questions. When they were asked what kind of recreational facilities the new town should have, the responses included a bowling alley, a movie theater, tennis courts, a golf course, an auto racetrack, handball courts and a roller-skating rink. One person suggested a burlesque hall.

Responses to the questionnaire reflected the independent spirit typical of many Alaskans. Under "additional comments and suggestions," one resident had written: "I don't believe in a lot of zoning laws telling me what to do on my land. If I like them I would of [*sic*] stayed Outside. I believe in minding my own business and others minding there's [*sic*]."

Some people in Valdez assumed that the new town would simply re-create the old; if they had a home or a business it would be on the same street or at the same intersection it had been in old Valdez. Finfer had other plans—and what planner wouldn't, given the haphazard nature of the old town's development?

But Finfer had very specific ideas about the way a town should be. He had studied architecture and planning at the Illinois Institute of Technology under Ludwig Hilberseimer, a German American who was an early proponent of the idea of "street hierarchy," in which a community's roads are classified by importance and function according to strict rules—so that, for instance, local roads don't connect with higher-traffic thoroughfares. Finfer had adopted the idea. For him, the pedestrian was king; a community needed to be designed so that home, school, work and shopping were all in close proximity, and children, especially, could walk to school protected from traffic.

Finfer held a series of public meetings to talk about his plan, which included residential cul-de-sacs and a car-free business district. Some in Valdez thought the plan was crazy for one reason—where would they put the snow? The town got upwards of twenty feet of it every winter, and when the streets were plowed there had to be places to put it. Finfer's design had made no allowance for that. As one resident put it later, the design "was more like an artists' colony where they don't have any snow." The plan for the business district was changed, with much wider streets and parking lots that in winter would hold snow rather than cars.

Finfer's plan was approved that summer. But doing the detailed design work, preparing the land, and building roads, water and sewer lines, docks and other infrastructure would take time, particularly since little could be done in the harsh winter months.

For home owners, the process would be arduous as well. Anyone who thought that Owen Meals's donation of much of the land meant that relocation would be without cost was in for a rude awakening. The state would buy home owners' houses, after an appraisal process. There would be an auction for lots in the new town, and home owners would have to arrange financing or otherwise figure out how to pay for a lot and for a new house if they chose to build one. If their old house was in good enough shape, they could choose not to sell it to the state and instead move it over to the new town. The whole process would take time, as there were plenty of bureaucratic hurdles to be overcome.

Home owners were dissatisfied with the appraisals of their current houses—though since the town had been through a major earthquake it's not difficult to understand why they would have relatively little value. But beyond that, most people in town had built their houses themselves or otherwise paid little for them, and if they'd ever had a mortgage they had long since paid it off. There were federal loans available for the move, and in some cases the payments could be deferred for a few years, but eventually home owners would face large monthly payments where they hadn't before. Valdez had hardly been a boomtown prior to the earthquake. Who knew if the new town would thrive economically and if its residents would do well enough to afford their new payments?

By the end of that first summer, Valdez residents could be forgiven if they felt unsure about their future. Current circumstances did little to lighten the mood. The Army Corps of Engineers had installed a temporary aboveground water supply system and had made repairs to the sewer system, so at least residents didn't have to leave tanks outside their house for pickup each day. But the water system was fragile—for some reason, the corps had used soft aluminum pipes, which kept getting crushed by cars. At one point, the town council had to threaten home owners with a fine of one hundred dollars if the pipes in front of their houses were damaged.

Worse, though, was that the tides now came into Valdez much farther than before the quake. The shaking that had caused the waterfront to collapse had had an effect on the sediments that the rest of the town was built on too. They didn't collapse, but they sank several feet.

The result was that some homes close to the water, even if they had made it through the quake, were now uninhabitable because of constant flooding. Walt and Gloria Day's house on McKinley Street was among them; they moved into one of the Salvation Army–supplied trailers.

Valdez wasn't such a fun place to live anymore, and it would be at least a year, and probably longer, before anyone could move to the new town site.

There had been a question of what to do about the schools. Valdez High School could be made usable for now. But the elementary school, even if it could be fixed, was judged a fire hazard. A new school was needed immediately. After months of being, as the *Bugle* had put it, "a town without children," Valdez was now starting to welcome back families with their youngsters. The town council realized that they could raise morale and show residents that a move was really happening by building the elementary school at the new town site.

On the face of it, it was a crazy idea. Design of the streets and utilities in the new town hadn't even begun, so the school's foundation would have to be built in such a way that it could be raised if needed to drain properly into the future sewer system. And transporting children every day to and from old Valdez to the new school would be costly.

But the council approved the idea, and a contract to prepare land for the school was signed on August 17. Prefabricated structures arrived by ship a week later. Then, one morning in the last week in September, the children of Valdez were taken by bus down the newly extended Richardson Highway to their six-classroom school. In some ways, they were the first residents of new Valdez. Over the next few years, they would be followed by everyone else.

13

DEEP THINKING

The revelation came one day near the end of the summer of 1964, in the backseat of a car in downtown Anchorage. George Plafker had been spending the summer in and around Prince William Sound, mapping the land changes caused by the Good Friday earthquake. Soon he would be heading home to California to begin the task of analyzing all the data he and others had collected, trying to make sense of it. Right now he was back in Anchorage for a bit, in a car with other scientists from the Geological Survey, talking shop.

Plafker had begun to get a clearer picture of what had happened in Alaska on March 27, but many things were still a mystery. He thought he had some understanding of why so much of the land affected by the earthquake had been uplifted, but he had more difficulty comprehending why so much land had subsided, or sunk. He'd seen this subsidence up close, in places like Girdwood, a small town along Turnagain Arm south of Anchorage, where trees near the shore that had been well above the high-tide line were now inundated when the tide came in. The salt water was killing them—and it was also killing Girdwood, which was now chronically flooding and would have to be relocated inland. Elsewhere, throughout northern and western parts of the sound, he'd precisely measured this subsidence, using the height of the barnacle line on the rocks as a reference.

As he, Arthur Grantz and Reuben Kachadoorian had written

in their report published a month after the quake, there was a clear "hinge" line, a zone of little or no elevation change between the areas of uplift and subsidence. But Plafker had learned other interesting things since that first intense month. For one thing, most of the aftershocks that had occurred since March 27—and there were thousands of them—had their epicenters within the zone of uplifted land. That was an indication that the main earthquake fault itself—the fault that must exist but that no one had seen evidence of on the surface—was beneath the zone of uplift and not beneath the subsidence zone. Or as Plafker was to put it later, the subsidence was "secondary" to the faulting.

Second, in addition to the work by Plafker and others at the Survey to measure uplift and subsidence, the Coast and Geodetic Survey had begun measuring horizontal movement—how much the land had shifted laterally during the quake. This was apart from any local movement, like the Turnagain Heights slide in Anchorage. Surveyors had found, for instance, that Montague Island, at the southeastern limit of the sound, had moved about sixty feet to the southeast. That was about the maximum extent of horizontal movement. Whittier had moved about twenty-five feet, and farther to the northwest the numbers had become smaller, until there was no horizontal movement at all on a line that began north of Anchorage and ran east to Glennallen.

Plafker had absorbed all of this information and turned it over in his mind many times. He could understand that when the fault had ruptured, the land over it had moved up and laterally. But he couldn't understand what had happened to the land that was *not* over the fault. It had moved laterally, too, though not as much. Why did it sink?

When he was alone with his thoughts for a moment in the backseat of the car, it hit him. The sunken land had, in a sense, been along for the ride. It hadn't moved so much as it had been *pulled,* from one side, by the uplifted land as it was moving laterally to the southeast. What happens when something is pulled like that? It stretches and thins out. It would be like holding on to a thick disk of pizza dough with one hand and pulling with the other—the dough would become thinner. That's what had happened here, Plafker thought. The land had been stretched and become thinner, and thinner land was lower land.

Plafker thought he now had the answer to why so much of the land had sunk. It would take a while for him to quantify this and state it in scientific language, and he'd need expert help. But it was only one of many points he'd have to study and explain, and he was excited about the prospect. While many scientists had viewed the earthquake as an interruption and had longed to get back to the work they were doing beforehand, Plafker had a different attitude. To him the quake was less of an annoyance and more of an opportunity, a chance to learn about concepts and principles that were far beyond what he'd needed to know in his studies and career so far. As he prepared to head back to the USGS's offices in Menlo Park, he knew there would be a steep learning curve. He would have to write a report for the Survey on the tectonics of the quake, one of many studies that were being produced. And no doubt he'd have to write a paper, too, for a professional journal. He'd be busy.

By the time of the 1964 earthquake, a little more than half a century had passed since Alfred Wegener first proposed his theory of continental drift. Wegener's idea, altered and adapted over the years, had made inroads among scientists, especially those in Europe. Harry Hess's hypothesis about seafloor spreading, as modified by Robert Dietz, had helped a lot. The "conveyor belt" concept—that oceanic crust formed when hot magma welled up at a midocean ridge, spread from the ridge as more crust formed, eventually sank when it met lighter continental crust and then looped back toward the ridge—provided something that Wegener lacked, a plausible mechanism by which the continents moved apart. Seafloor spreading also helped explain why to that point no oceanic crust had been found to be more than about 150 million years old: by Hess's thinking, it was recycled back into the mantle before it got any older. The hypothesis also provided a mechanism for mountain building: as the oceanic crust reached the end of the conveyor belt and sank down into deep trenches, sediment that had accumulated on it during the tens of millions of years while it was moving would be scraped off by the continental crust, lifted up and, over more time, become mountains.

In the mid-1960s, however, there were still plenty of "stabilists" around. Some accepted the concept that there was upwelling of magma at the ridges but had a much harder time with the idea that the oceanic crust was recycled at the continental margins. Instead, they thought that the magma just added to the earth's crust, making the planet bigger. This "expansion theory" had first been proposed in the 1930s and, through various permutations, still had adherents.

In early 1963, a young English graduate student at the University of Cambridge came up with an idea that was, as he described it later with typical British understatement, a "rider" to Hess's hypothesis. The student, Frederick Vine, had been inspired to study earth science eight years before when, as a fifteen-year-old in West London, he'd cracked open a geography textbook while studying for his O-level exams and saw a diagram illustrating how well the coasts of South America and Africa fit together. It was the same congruence of coastlines that had inspired Wegener. In reading the text that accompanied the illustration, Vine had learned that although there was a theory that explained this close fit by proposing that the continents had once all been together but had since come apart, geologists had no idea if it was true. Vine thought it remarkable that something as basic as this was not known. He went on to study geology at Cambridge as an undergraduate and in early 1962 attended a conference at the school at which Harry Hess, as guest speaker, gave a talk on seafloor spreading. Vine was hooked on the idea of trying to prove that the hypothesis was correct.

In the graduate program, Vine was assigned to work with a geology professor, Drummond Matthews. One of Matthews's responsibilities was to obtain geomagnetic data; Vine's job, as his research assistant, was to help interpret it. Beginning in the 1950s, scientists at several institutions, including Lamont Geological Observatory in New York, had been collecting data about the magnetism of the seafloor. They relied on a fact about rocks that had been known for centuries: many, notably volcanic ones that contain a lot of iron, are magnetic. They become magnetized when they form, cooling from extreme heat, and groups of atoms within them line up with the earth's magnetic field. As the rocks harden, that alignment is locked in. The rocks then contain a permanent record of the orientation of the magnetic field—of the direction to the magnetic pole—and the field's intensity at the time the rocks formed. This can be measured using an instrument called a magnetometer.

Researchers in the field of what became known as paleomagnetism had been studying rocks collected on land in this way, using lab-based magnetometers. In the mid-1950s, Keith Runcorn, a physicist at Newcastle University in England, looked at data from rocks from different time periods in Europe and concluded that the magnetic pole had migrated over millions of years—what was called polar wandering. Later, he compared that data with similar studies of North American rocks and realized that a better explanation was not so much that the magnetic pole had wandered, but that the rocks—and thus the continents themselves—had moved. This spurred renewed interest in Wegener's concept of continental drift among Runcorn and others. Eventually, with enough data from rocks around the world, these scientists were able to accurately infer how the continents had shifted in relation to one another, findings that added to the growing evidence that Wegener had been onto something.

To collect data from the seafloor, scientists from Lamont and elsewhere trailed a magnetometer off the back of a ship as the vessel tracked back and forth across a survey area. The magnetic maps that resulted were intriguing, as some areas showed stronger magnetism than others. Often these formed a pattern—narrow strips of the seafloor, alternating strong and weak, like the stripes of a crosswalk.

Scientists studying these maps were not sure what the patterns represented. They had known for some time that rocks can exhibit what is called reverse polarity. That is, the locked-in magnetism in some rocks is oriented one way; in others it is aligned in the opposite direction. Hypotheses had been proposed to account for this, including what seemed like a radical one: that from time to time the earth's magnetic field flips, and that the magnetic pole, what we call magnetic north, becomes magnetic south. Then at some point it flips back again. This was much more than polar wandering; this was polar reversal. Rocks that formed during a period when the magnetic field was oriented as it is today would have one kind of polarity. Those that formed when the field had flipped would have reverse polarity.

The idea that the earth's magnetic field reverses was not widely accepted—for one thing, no one had a satisfactory explanation as to *why* it would happen—but it began to gain more credence as more data suggested that the field had flipped often over millions of years. But it was still a contentious issue when Vine arrived for graduate study in the fall of 1962. As it happened, Matthews was away on a research cruise aboard the HMS *Owen,* an old Royal Navy minesweeper, conducting magnetic surveys across a stretch of the Indian Ocean that included a midocean ridge. When Matthews came back later in the fall, Vine looked at the data. Like that from other surveys, it contained patterns of anomalous readings.

Vine had a thought. What if these readings represented areas of alternating polarity, polarity that was locked in as—following Hess's reasoning—magma oozed out of the midocean ridge and cooled to become new seafloor? As seafloor kept being produced over millions of years and moved toward the continental margin, reversals of the earth's magnetic field would create stripes of seafloor of alternating polarity. That would account for the crosswalk-like appearance of the magnetic maps.

In a paper published in 1963, what came to be known as the Vine-Matthews-Morley hypothesis (recognizing that a Canadian scientist, Lawrence Morley, had come up with the idea independently at roughly the same time) offered a neat way to prove the seafloor-spreading hypothesis. The data that Vine and Matthews offered wasn't clear-cut, so the stabilists, especially, were skeptical. But if seafloor spreading actually occurred, further analysis of magnetic data—and by then there was a lot of it—should show stripes of seafloor of alternating polarities, and these stripes should be roughly symmetrical on either side of a midocean ridge. What's more, analysis of the age of rock samples from the seafloor should show the stripes becoming progressively older the farther they are from the ridge.

By the fall of 1964, when George Plafker started working on his data in earnest in his office in Menlo Park, no such proof of the Vine-Matthews-Morley hypothesis—and thus of Hess's seafloor-spreading hypothesis—had been published. But Plafker wasn't waiting for proof. By then, he was becoming something more than a field geologist. In the months after the earthquake he had learned a lot of geophysics—in particular, seismology—and he would learn a lot more. He was aware of the disagreements between stabilists and mobilists and found the arguments put forward by the mobilists compelling. But he also had an intuitive sense that they were right. He kept going back to his time in Bolivia and the dropstones he'd seen on the eastern side of the Andes. He knew they had to have been deposited by glaciers, and it seemed to him that the glaciers could have come from nowhere else but another continent—one that had once been attached to South America but had long since drifted away.

Plenty of scientists weren't convinced that Hess and others were right, especially in the United States. Lamont Geological Observatory had a number of researchers who were skeptical of "spreaders and drifters," as the mobilists were sometimes called, even though the lab had produced much of the data regarding the physical and magnetic characteristics of the ocean floors. Lamont's director, Maurice Ewing, once asked a colleague before a talk about seafloor spreading, "You don't believe all this rubbish, do you?" and as is not uncommon in science, some lower-level researchers took their cues from those above them. There were also stabilists at Menlo Park, geologists who took a rather narrow view: they didn't see evidence for the movement of continents in their own work and thus rejected the idea.

But some noted researchers at the Survey were firmly in the mobilist camp, and Plafker was fortunate that one occupied a nearby office. His name was Allan V. Cox.

Cox wasn't just any researcher. Just three years older than Plafker, already by 1964 he was one of the foremost experts on paleomagnetism in the world, one whose work had been cited by Vine and Matthews in their paper the year before. A native Californian, Cox had dropped out of the University of California at Berkeley early during his freshman year, had served in the merchant marine for several years, had worked as a field assistant studying glaciers in Alaska and then had been drafted into the army. Upon being discharged he returned to Berkeley, eventually graduating with a degree in geology. After more time in Alaska, in the mid-1950s he came back to Berkeley once again for his doctorate, studying the magnetism of rocks under a faculty adviser who was a firm believer in continental drift. In 1959, with his PhD in hand, he'd been hired at the Geological Survey. Cox, with another Survey scientist, Richard Doell, quickly waded into the subject of magnetic field reversals. Their idea was to look at magnetized rocks around the world and determine their ages using the latest dating techniques, to put together a timeline for when the magnetic field had reversed.

It was an ambitious undertaking, involving a lot of fieldwork (and they had to persuade the Survey to hire another scientist who was an expert in dating). But they'd published a paper in 1963 that established a partial timeline of reversals—a proof of concept, as it were—based on analysis of samples from six lava flows in California.

Plafker was familiar with their work. In fact, he had been responsible for inadvertently ruining Doell's summer a couple of years before, when Doell had come to Alaska to find rocks for his magnetic studies. Plafker was helping a colleague map rock formations in the Wrangell Mountains with an eye on potential deposits of copper ore, and Doell had worked out of their camp. There were layers of volcanic rock in the area dating back to the Permian period, more than 250 million years ago, which would be perfect for his research. He'd spent weeks collecting samples while Plafker and his colleague were doing their mapping work. Then, toward the end of the summer, Plafker came across a layer of shale-like rock below the volcanic layers. In looking at the shale closely, he noticed it contained fossil shells that were easily recognizable as a certain kind of extinct clam—but one from the Triassic period, less than 250 million years ago. If the shale was less than 250 million years old, that meant that Doell's volcanic layers, above the shale, must be younger, and not from the Permian, as he had thought. Thanks to Plafker's find, weeks of Doell's work went out the window.

In the fall of 1964, Cox was still in the thick of his paleomagnetism research for the Survey, though in a few years he would leave for a professorship at Stanford, where he'd work for two decades until his death in 1987. Plafker thought Cox was brilliant and a bigger thinker than most of the other scientists at the Survey, who seldom looked much beyond the quadrangle they were mapping or the basin they were studying. Plafker would go to Cox's office for brief conversations about what he was finding in the data, the conclusions he was coming to. Cox was an unabashed mobilist and offered Plafker encouragement. So did George Gryc, the Alaska branch chief who had sent him to Anchorage after the earthquake. Gryc thought it great that Plafker had thrown himself into this work, when so many others would have complained that it was a career diversion.

Plafker's task was essentially this: how to describe what had happened beneath Alaska beginning at 5:36 p.m. on March 27 in a way that best accounted for the observed effects, primarily the changes in land level around the state. What mechanism would cause such massive uplift and subsidence over tens of thousands of square miles of land?

The key, he knew, was to understand the major fault. The two faults he'd seen on Montague Island had to be only secondary features. But where precisely was this main fault, how big was it and, most important, how was it oriented? He'd seen no evidence of it on the land surface, but there were clues in the data.

From what Plafker knew of seismology, his choices of a type of fault were limited. Faults can be complicated, but they can be divided into three basic categories:

Strike-slip, *in which one side, or block, slips past the other horizontally.*
Dip-slip, *in which one block moves past the other vertically, or nearly so.*
Thrust, *in which one block rides over the other at a low angle.*

In the case of the Alaska earthquake, Plafker could rule out a strike-slip fault. It was extremely difficult to see how rocks sliding against each other horizontally could cause either side to rise up or sink, especially by the amounts measured over such a wide area of south-central Alaska.

As for the other two types of faults, "first-motion" studies provided some clues. A first-motion study analyzes the waves from an earthquake to determine the fault plane, the angle that the fault makes in the earth. The problem with these kinds of studies is that they normally produce two possible solutions rather than one. With the Good Friday quake, the analysis suggested that the fault could be at a steep, nearly vertical angle—a dip-slip fault—or at a low, nearly horizontal one—a thrust fault.

Plafker knew that a well-known geoscientist—no less than the head of the seismology lab at the California Institute of Technology, in Pasadena—was also working on the earthquake and had determined that the fault must be a steep, dip-slip fault. In the geology and geosciences community, as in most scientific communities, word gets around as to what people are working on. But Plafker also knew this because the scientist Frank Press had asked Plafker for his data, which he had shared.

Five years older than Plafker, Press was, like him, a native New

Yorker who had graduated from City College. But unlike Plafker, who had found a job after graduation and headed west, Press had continued his education at Columbia, getting a doctorate in geophysics in 1949. His adviser had been Maurice Ewing, who guided Press in designing an improved type of seismograph. Press remained at Columbia until 1955, when he moved to Caltech, which at the time was considered the world's leading institution for research in seismology. Two years later he was named director of the Seismological Laboratory, a somewhat surprising choice given his relative youth. But Press, while an extraordinary scientist—among many accomplishments, he was at the forefront in understanding the seismic signatures of earthquakes versus atomic bomb tests and had helped design the worldwide network of seismographs that measured the Good Friday earthquake—also had a head for policy making. In the early 1960s he had served on the President's Science Advisory Committee and had been a delegate to international meetings in Geneva and Moscow on limiting nuclear weapons testing. In 1962 and 1963 he was president of the Seismological Society of America.

Although Plafker had readily shared his data with Press—it was the government's, after all—as he'd heard through the grapevine about Press's conclusions, he found them difficult to accept. How could a near-vertical fault produce the kind of large elevation changes on either side of it that had been seen across such a huge swath of south-central Alaska? To Plafker, it couldn't. If a dip-slip fault had had such a widespread effect, surely there would be some sign of the fault on the surface, like a scarp where one side of the fault was now higher than the other.

But Press was an important and highly respected scientist, and in the groupthink that sometimes affects science, plenty of lesser scientists felt that his explanation must be the correct one. Plafker, an unsung field geologist who became involved only because of his knowledge of Alaska, who until a few months before had known almost nothing about seismology, felt trepidation going against such an authority. But he was slowly figuring out a mechanism for the earthquake that differed from Press's and, with the encouragement of Cox and others, felt his solution was correct.

Plafker thought the alternative fault orientation—a fault that dipped at a low angle to the northwest—was the right one for two main reasons.

One had to do with the aftershocks. As he'd learned before, the vast majority of these occurred within the land that had uplifted. But the depth of the strongest aftershocks—about 130 that were of Richter magnitude 5.0 or higher—was even more of a clue. These tended to be deeper the closer they were to the hinge line. That tendency would be expected if, as Plafker thought, the fault was dipping to the northwest at a low angle, toward the hinge line.

But the other reason had more to do with what Plafker had come to learn, and accept, about seafloor spreading. If crust that made up the floor of the Pacific Ocean was sliding and sinking beneath the crust that made up Alaska and the rest of the North American continent, it wasn't hard to envision that this was happening at a shallow angle. There'd be no reason to think, for example, that the heavier ocean crust would dive sharply down when it met the continent. Rather, a slow, gentler, more gradual sinking made intuitive sense.

So to Plafker's thinking, the earthquake fault was a thrust, not a dip-slip, fault. It occurred where the oceanic crust and continental crust were in contact, as the former was sliding, or subducting, under the latter. Given that the fault was nearly horizontal, the fact that no surface trace of it had been discovered now made sense.

According to the distribution of aftershocks, the fault was about five hundred miles long from the northeast to the southwest and more than one hundred miles wide from the southeast to the northwest. This made it a huge fault. But given the size of the earthquake, it *had* to be huge. Energy had to have been building up, for a long time, along the fault before it slipped on March 27.

How had the energy built up? As the oceanic crust slid underneath the continental crust, friction in the zone where the two met would cause the continental crust to be compressed. It would be as if the oceanic crust were tugging on the continental crust and squeezing it. Plafker thought he had seen evidence for this on Middleton Island. He'd seen young rocks that had clearly been folded and otherwise deformed, indications that they had been under compression. The earthquake had occurred when the strain between the two types of crust became so great as to overcome the friction between them. This resulted in the sudden release of an enormous amount of stored-up energy over the wide area of the fault.

Plafker had his fault, and he had his earthquake mechanism. It all made sense to him. It explained both the uplift and the horizontal movement: when the fault ruptured, the continental crust rebounded like a spring, up and out, in the opposite direction from the compression. This crust movement also explained the secondary faults on Montague Island, which came to be called splay faults (and were later determined to be the cause of some of the destructive tidal waves). And as he had figured out before, the movement explained the subsidence that occurred to the northwest of the uplift zone.

This was no ordinary thrust fault: it was so big, and produced such enormous effects when it ruptured, that just calling it one seemed inadequate. It would eventually come to be known as a "megathrust" fault.

Press's paper, written with a Caltech undergraduate, David Jackson, appeared in *Science*—a publication of the American Association for the Advancement of Science and the most prestigious general science journal in the country—on February 19, 1965. Plafker had been generally aware of the findings, so he wasn't shocked to see them in print. Yet he still couldn't understand why Press had come to the conclusions he had.

According to their analysis, a near-vertical dip-slip fault ran southwest to northeast for close to five hundred miles along the hinge line. It began, they calculated, about 10 miles below the surface. In itself that wasn't unreasonable, Plafker thought, and it would explain why there was no surface evidence of the fault. But Press and Jackson proposed that their fault extended to depths of 60 to 120 miles. This, as they acknowledged in the paper, was an order of magnitude deeper than any known earthquake fault.

To Plafker, the extreme depth should have been a clue that perhaps their fault concept was incorrect. For one thing, the data showed that the deepest aftershocks originated only about 25 miles down. If the fault was as deep as Press and Jackson said it was, some of the aftershocks should have occurred at greater depths.

Plafker kept working on his ideas, convinced more than ever that he was on the right track. He began writing what would turn out to be, by the standards of academic journals, a long paper—more than three thousand words, with numerous photographs and diagrams—and submitted it to *Science.*

The work, "Tectonic Deformation Associated with the 1964 Alaskan Earthquake," was published on June 25, four months after Press's paper. Plafker was the sole author, and he laid out the evidence he and others had collected, most notably the more than eight hundred measurements of uplift and subsidence. He detailed the use of barnacles and other marine organisms as reference marks. He wrote about the large tidal waves and what was known about their cause, and about the secondary faults that had been discovered on Montague Island.

The last page and a half of the thirteen-page paper was devoted to discussing the main fault and other aspects of how the quake originated. Quietly, with the kind of dry, clinical language befitting a scholarly paper, Plafker dismantled Press and Jackson's arguments. In describing the changes in land level he and others had recorded, he noted that "these measurements showed no abrupt changes of level indicative of vertical fault displacement." In discussing the idea that, of the two possible solutions, a near-vertical fault was the correct one, he wrote that "the hypothesis appears to pose more problems than it answers."

The most serious of these problems, Plafker continued, was the absence of any surface displacement at the fault line. There was no scarp or other feature in the earth where the fault was presumed to be, but there was plenty of displacement to either side of it.

He then pointed out that Press and Jackson, "in an elegant analysis of the displacements," had shown that the uplift and subsidence could be accounted for by a fault that began ten miles below the surface. This was the extra-deep fault that to Plafker seemed unbelievable. He cited four "serious objections" to it:

1. A near-vertical fault of this size and destructiveness should have broken the surface somewhere.
2. It should have produced land-level changes opposite of what actually occurred. The land to the northwest should have been uplifted, and that to the southeast should have subsided.
3. It was far deeper than any known fault.
4. Aftershocks occurred on only one side of it.

In place of Press and Jackson's hypothesis, Plafker laid out his own, that the fault in question was a low-angle thrust fault. He also described the mechanism by which energy had been stored in it: "The postulated stress pattern could result from progressive underthrusting of the oceanic crust and mantle beneath the continental margin."

The main weakness of the thrust-fault argument, he pointed out, was that while it could account for the uplift and the seaward lateral movement, it had a harder time accounting for the subsidence. But Plafker detailed his idea that the subsidence might be "a secondary effect resulting from elastic deformation" when the land was pulled and stretched.

Toward the end of the paper Plafker trod lightly, describing his findings as speculation and noting that much more research was needed. But he also left little doubt that he believed his ideas about the fault and the origin of the earthquake were correct

Plafker had had two things going for him, he thought. One was that he had seen the effects of the earthquake for himself.

He strongly believed in the importance of observation; there was no substitute for it. He had practically lived and breathed the earthquake for months, traveling all over the affected area, talking to people who had lived through it and measuring the land level changes himself. Before that, earlier in his career, he'd seen things in the field, notably in Bolivia, that had informed his thinking as well. So when he started analyzing the data, he had an intuitive understanding of what made sense and what didn't. Press and Jackson's solution just didn't fit with what he'd seen and experienced.

Second, Plafker was convinced that seafloor spreading and its related concepts were real. He felt that his findings fit well with those concepts, certainly much better than Press's did. A near-horizontal thrust fault fit much more easily with the idea of oceanic crust sliding underneath continental crust than did the near-vertical fault that Press proposed. But Plafker's conviction was stronger than that. As he saw it, the only way to understand what happened in south-central Alaska on March 27, 1964, was to accept the ideas that had originated with Alfred Wegener half a century before and had been altered and adapted by the likes of Harry Hess, Fred Vine and others.

They were ideas that would soon be further altered and adapted and supported by the work of scientists around the world. In a few years—thanks in part to Plafker's work—they would become fully accepted, and much better known, as the theory of plate tectonics.

14

ACCEPTANCE

The invitation was a little unusual. It was the spring of 1967, and George Plafker was being asked to make a presentation at one of the biggest scientific conferences of the year—of geophysicists. The forty-eighth annual meeting of the American Geophysical Union was to be held in Washington in mid-April, over four days at the Sheraton-Park Hotel, the sprawling convention facility overlooking Rock Creek Park. It was not lost on Plafker that he, a geologist who spent much of his time in the field looking at rocks, would be speaking at a session with nine other scientists who spent much of their time at a desk looking at plots of seismic or magnetic data. Among those nine were some of the world's leading thinkers on the subject of the workings of the earth's crust, including the man responsible for the theory of seafloor spreading, Harry Hess. Plafker might feel a little out of his league—he would say later that he was the "token" geologist among the presenters—but he was flattered that the geophysicists thought he had something important to say.

His paper in the journal *Science* hadn't exactly caused a sensation when it was published in 1965. Not that most people expected it would, but even those who might have been disposed to see great significance in it—after all, the paper quietly implied that the Alaska earthquake was a real-world validation of the ideas that

Alfred Wegener and Hess and others had talked about all these years—were busy looking for their own proof of those ideas.

The publication in 1963 by Fred Vine and Drummond Matthews of their hypothesis that oceanic crust consists of sections, or stripes, of alternating magnetic polarity had set off a rush to find data that confirmed it. Much of that data was already available, and much of it had been obtained by researchers at the Lamont lab, where being a nonbeliever in seafloor spreading had not prevented the director, Maurice Ewing, from establishing a vigorous program to conduct magnetic surveys of the oceans. Analysis of the data over the next few years led to a sea change, of sorts, at Lamont, with more scientists joining the mobilist camp. Even Ewing showed signs of softening his position. By 1967, several papers by Lamont researchers had been published that validated the hypothesis.

Besides paleomagnetism, research in other fields contributed to growing acceptance of the seafloor-spreading idea. Another Lamont researcher, Lynn Sykes, had studied earthquake data to better characterize the midocean ridges. Bryan Isacks and Jack Oliver, both from Lamont as well, reported on evidence for subduction in seismological data from the vicinity of Tonga, in the South Pacific. In 1965, J. Tuzo Wilson, a geophysicist at the University of Toronto, proposed the idea of transform faults, between segments of the midocean ridges, recognizing that crust could sometimes move in ways that didn't result in its being created or consumed. And in 1966, Dan McKenzie of the University of Cambridge applied thermodynamic principles to studying the mantle and came up with a more detailed mechanism of crust movement.

With momentum growing, the Geophysical Union decided to devote a full day of sessions to seafloor spreading and related subjects at its 1967 meeting. But there were many other subjects on the program: the conference included a who's who of geoscience delivering talks on cutting-edge research in fields like meteorology, hydrology, volcanology and planetary science. Among those giving presentations was Frank Press, who had moved from Caltech to the Massachusetts Institute of Technology and had turned some of his attention to the moon. He was on the bill at two sessions on lunar seismicity on Monday and Tuesday, including one with his old adviser Maurice Ewing of Lamont.

The sessions on seafloor spreading were scheduled for Wednesday. The morning session included talks about seafloor magnetics, sediments, gravity studies and other subjects. Plafker's presentation was one of ten in the afternoon. He would go fourth, and like the others among the first nine speakers would have twenty minutes. Hess, whose ideas were the star of the show, would go last and have thirty minutes.

Plafker's talk was titled "Possible Evidence for Downward-Directed Mantle Convection Beneath the Eastern End of the Aleutian Arc." It was a further discussion of what he'd stated in his 1965 paper, and although the title used the word *possible,* Plafker left no doubt about what he believed, that the Alaska earthquake was evidence of "downward directed mantle convection," or subduction, of oceanic crust.

Press, who was finished with his own presentation duties, was in the audience for Plafker's talk. Afterward, Press approached him and, by Plafker's recollection, acknowledged that his own 1965 paper proposing a vertical earthquake fault had been wrong and that Plafker's analysis had been correct. Plafker was stunned. Here was one of the most renowned scientists in the country telling him—a geologist among geoscientists—that he'd gotten it right.

Press wasn't the only distinguished researcher to approach Plafker after his talk. Clarence Allen, who had taken over as interim director of the Caltech Seismological Laboratory after Press left for MIT, came up to Plafker with a suggestion: Why don't you look into the 1960 earthquake that struck the coast of Chile? The conditions there appeared to be similar to Alaska: oceanic crust colliding with continental crust, in this case South America. If your ideas about the Alaska quake are correct, Allen told Plafker, you should find confirmation in Chile.

The earthquake, which had struck about one hundred miles offshore in the midafternoon of May 22, 1960, was generally considered the strongest ever recorded. It was centered about 350 miles from the capital, Santiago; the nearest city of any size was Valdivia, and it was devastated.

The Chile quake was like the Alaska one, only more so. The shaking lasted longer (ten minutes by some estimates), more houses were destroyed (thousands in Valdivia alone) and more and larger tidal waves did even more damage, both in Chile and hours later across the Pacific. A wave estimated at thirty-five feet hit Hilo, Hawaii, killing sixty-one, and on the Japanese islands of Honshu and Hokkaido nearly two hundred people died. The death toll in Chile has never been ascertained, but at least 1,600 people, and probably more, were killed—as in Alaska, mostly by water. Even in 1967, some parts of coastal Chile were still struggling to recover.

Plafker had leaped at the idea of studying the quake, and Allen had the power to help him obtain a grant for the work so he could take time off from the Geological Survey. With funding secured, he went to Chile in 1968, at first talking to fishermen along the coast about tidal changes there. His Spanish, still impeccable from his time in Guatemala and Bolivia, proved invaluable. Then he'd chartered a fishing boat, the *Atun,* with a Chilean crew, to measure uplift and subsidence directly among the many small islands that dot the Chilean coast. The work was not as straightforward as that in Alaska—much of the elevation shifting had occurred underwater—but in a paper published in 1970, Plafker described the same basic changes he'd seen in Alaska. There was a zone of uplift to the west and one of subsidence to the east that together encompassed a huge area. He identified a "zone of faulting" that ran roughly six hundred miles north to south and was at least thirty-five miles wide. As in Alaska, he identified the fault as a thrust fault dipping only slightly from horizontal (although the direction was different because off Chile the seafloor was moving almost due east, not to the northwest as in Alaska). As Clarence Allen had suspected, the Chile quake was essentially the same as the Alaska one: the result of oceanic crust sliding, or subducting, beneath continental crust.

Plafker and a colleague, James Savage, wrote a paper describing the findings. By the time it was published in the *Bulletin of the Geological Society of America,* more scientists were becoming comfortable with referring to those sections of crust as "plates." The ideas of Wegener, Hess and others were increasingly being accepted and combined into what was becoming known as the theory of plate tectonics.

One of the talks given at the American Geophysical Union meeting in 1967—one that was somewhat overlooked at the time—had played a role in this increasing acceptance. It was given during the morning session by Jason Morgan, a young geophysicist from Princeton, and was the last talk before the noon break. Because of the timing, many in the audience left before he started, wanting to beat the long lines at lunch. Morgan also changed the subject of his talk—rather than discussing how ocean trenches were formed, the topic that had been published in the meeting program, he announced that it would instead be about "rises, trenches, great faults and crustal blocks." Over the next twenty minutes, Morgan proceeded to lay out, essentially, the theory of plate tectonics. This was believed to be the first time that the term *plate tectonics* was used in a public setting.

Morgan's was a work of synthesis, which he described in greater detail in a paper submitted in August 1967 and published in 1968. Dan McKenzie, who had moved to the University of California at San Diego, and a colleague, Robert Parker, published a paper in December 1968 describing the same general ideas.

Simply put, the theory holds that the outer layer of the earth consists of a number of rigid sections, or plates—scientists at first identified about a dozen large ones, though smaller ones were discovered later—that are in constant slow motion with respect to one another. Upwelling of hot magma at ridges is the engine that drives the motion.

It's not that simple, of course. There would still be plenty of work to do to flesh out the theory—indeed, plenty of work is still being done today. And it would be years before it was completely accepted, for all intents and purposes. But in its own way, the theory of plate tectonics is now considered as consequential as Darwin's theory of evolution (although plate tectonics was the work of many people, not one man). The theory shows that Alfred Wegener's basic idea, that the surface of the earth is dynamic, was correct, even if some of his specifics were wrong (for one thing, it's not just the continents that are moving—all of the earth's crust is in motion). But plate tectonics explains much more than crustal movement. It is a unifying theory for all the geological features and processes that humans have wondered about for centuries; they can all be seen through its lens. It accounts for mountains and rift valleys, for volcanic eruptions, for hot spots and ocean trenches and tall undersea mountains.

And it accounts for earthquakes, too. Thanks to the understanding of plate tectonics, and the work of Plafker and many other scientists, science now has far greater knowledge of earthquake hazards around the world.

The 1964 quake, as Plafker described it in both the 1965 *Science* paper and a longer report for the Geological Survey published in 1969, was the first recognized "megathrust" earthquake. But beginning with Plafker's work on the 1960 Chilean quake, scientists began to see megathrust earthquakes everywhere. Subduction of denser oceanic crust—the Pacific plate—is occurring all around the "ring of fire" in the Pacific, and megathrust earthquakes occur in these subduction zones. The 2011 quake off Japan that led to the disaster at the Fukushima nuclear power plant is just one example of a recent major megathrust quake. But subduction-related quakes occur elsewhere around the planet too, even in places where the collision is between two continental plates, like the Himalayas, where an April 2015 earthquake killed more than nine thousand people in Nepal.

Even earthquakes that are not of the megathrust type, like the 1906 San Francisco quake and others that have occurred on the

San Andreas Fault, are understood to be related to plate movement. The mechanics are complicated, but the San Andreas, for example, is a transform, strike-slip fault between ridges; the slip movement is horizontal. The Fairweather Fault, which caused the 1958 quake that led to the giant wave in Lituya Bay, Alaska, is a transform fault as well.

The Alaska quake had other lasting impacts, too. That so much of the death and destruction was attributable to water led to extensive research on tsunamis (the Japanese word having eventually gained preference over the less-than-accurate term *tidal wave*) and what generates them. In large part because of what Plafker learned about uplift and subsidence in 1964 and the quake motion that he identified, scientists learned how to recognize areas of tsunami risk and engineers better understood how to design protective structures. Tsunami warning networks, installed in some parts of the world since the late 1940s, were expanded and improved.

The collapse of the Valdez waterfront and the slides in Anchorage were just two elements of the destruction in Alaska that revealed, to an extent seldom seen before, the dangers of soil liquefaction during an earthquake. Engineers learned that identifying soils at risk of failing in a quake was as important as designing structures to withstand shaking. Earthquake codes were made stricter, in Alaska and elsewhere.

For seismologists, the Alaska quake led to improvements in measuring the power of earthquakes. The 1964 quake—during which the closest seismograph essentially went off the charts—revealed some of the limitations of the Richter magnitude scale. The 1960 Chilean quake, too, had shown that the scale wasn't particularly accurate for large events. By the 1970s, scientists had developed a different approach—using what is known as "moment magnitude," which takes into account the size of the fault and the amount of slip and can be determined from the size of the seismic waves.

On this scale, the 1964 Alaska quake has been calculated at magnitude 9.2. That makes it the most powerful earthquake in North America ever measured by instruments, and the second most powerful—to the 1960 Chilean quake, calculated at 9.5—in the world. The 2011 Japanese earthquake was measured at 9.1, which means that the Alaska quake, while only one-tenth larger in magnitude, released about 50 percent more energy.

It's in helping scientists understand earthquake risks in other areas of the world that the Alaska quake has had its biggest impact. And among these other areas, none has a potentially graver risk than the Pacific Northwest—from British Columbia to Oregon.

There, off the coast, conditions are similar to those in Alaska. Rather than the large Pacific plate, however, three smaller oceanic plates are sliding under the North American continental plate in an area called the Cascadia subduction zone. Stresses are building up, just as they did in Alaska until March 27, 1964.

While scientists cannot predict earthquakes—although over the years many, including Frank Press, have tried to develop ways to do so—they are able to forecast, more generally, the earthquake risk in a certain location, especially if they know when earthquakes occurred there in the past. Once again, studies of the Alaska quake have been crucial for recognizing and understanding past earthquakes elsewhere, a field known as paleoseismology.

Scientists know that the last major earthquake in the Cascadia subduction zone was in 1700. The risk of another one is now considered high, with some parts of the coast estimated to have as much as a one-in-five chance of seeing a quake of magnitude 8.0 or higher in the next half century. Such a quake could spawn tsunamis and cause destruction along a coast that, with Seattle, Vancouver and other large cities, has a population close to ten million, or about forty times that of Alaska in 1964. Alaskans were caught unawares back then; but thanks to the scientific understanding gained from that event, today the people of the Pacific Northwest know they need to prepare.

15

EPILOGUE

To reach Alaganik Slough, you drive out of Cordova on a road to nowhere. State Route 10, also known as the Copper River Highway, curves past Lake Eyak with its aquamarine glacier-fed water, and heads east past the airport, named for Merle K. Smith, a pioneering bush pilot who had perhaps the ultimate Alaskan bush pilot nickname, Mudhole. You turn off for the slough at Mile 17 (the highway goes only to Mile 36, where the river overwhelmed a bridge several years ago; there are no plans to replace it) and drive south for a mile or so. This side road slices across meadows punctuated by ponds and willows, alders and small stands of Sitka spruce, their dark-green boughs standing out against the more muted yellows and browns that dominate the landscape.

The slough is on the western edge of the vast Copper River delta, where the river, on its way down from the Wrangell Range, separates into braids. These meander across the slightly sloping land, depositing huge volumes of glacial sediment before reaching the Gulf of Alaska. To the northeast, on the other side of Route 10, are the Chugach Mountains, with their snowy crests. Somewhere out there is Shattered Peak, the mountain that lost its top during the earthquake, causing a landslide that covered much of Sherman Glacier several miles away. A mile or so in the opposite direction is the gulf.

The day is calm and beautiful, unseasonably warm for spring, and quiet save for the drone of the occasional plane approaching the airport and the cawing of hundreds of gulls. The hooligan are running, and the gulls are circling over a spot in the slough's main channel where the fish are concentrated. The birds are anxious for a feast.

It is late April 2015, fifty-one years after the Good Friday earthquake. George Plafker, now eighty-six, and I have parked our rental SUV and tromped across the spongey peat-filled meadow toward the channel. We've come to the slough for a reason: Plafker needs a few more elevation measurements for research he has been doing regarding the history of great earthquakes in Alaska. A few years before, he took some cores around the slough; they are in cold storage back at the Geological Survey's office in California. I am here as his rodman, his field assistant, whose main job is to hold up a stadia rod so that he can make measurements by sighting on it from some distance with a hand lens. Plafker is tech-savvy and could be using some kind of digital surveying tool. But for this task that is overkill. I think he also wants me to get some small sense of what geology fieldwork is like. Being a rodman is, as Plafker has said more than once with a slight cackle in his voice, such an easy job that even I can do it.

Three years after his work in Chile in 1968, Plafker got his doctorate from Stanford, writing a thesis comparing the Chile and Alaska quakes. He spent years working further on the geology and tectonics of Alaska, but he also studied many other earthquakes and tsunamis. He returned to Guatemala to study the 1976 quake there that killed twenty-three thousand people. Years after the 1970 Ancash earthquake in Peru, he went there and analyzed a landslide caused by the quake. The slide was far bigger and worse than the one at Shattered Peak—the debris at one point went over a ridge and became airborne, reaching peak speeds estimated at 600 miles an hour and burying the town of Yungay. All but about one hundred of the twenty thousand residents were killed. Later, he traveled to Aceh Province in Indonesia to study the 2004 tsunamis that killed more than 130,000 people there (and which, he found, were caused by movement on splay faults). To determine how soon after the quake the first tsunami had arrived in Banda Aceh, the capital, he bought some broken wristwatches from a shopkeeper whose store had been all but destroyed. The watches had stopped when the water hit.

Plafker retired from the Geological Survey in 1995, but in reality he has never retired at all. As an emeritus geologist, he still has an office in Menlo Park—at the earthquake branch now, not the Alaska branch—and most days he makes the short drive there from his home a few towns away.

The Alaska earthquake changed the lives of all who lived through it that day in 1964. Plafker didn't experience the disaster—he arrived a day after the shaking stopped—but in his own way, and just as surely, the quake altered his life too.

The foothills of the Sierra Nevada in Northern California look something like the foothills of the Chugach in south-central Alaska, minus all the precipitation. There hasn't been much rain, or nearly enough snow, in these hills east of Sacramento for several years, so that by the summer of 2015, as I drove up a long hill on the outskirts of Placerville, the landscape was very dry and very brown. At the top of the hill, down a long driveway, was a comfortable one-story house on six acres. It's the house where Kris Madsen, now Kris Van Winkle, has lived for more than thirty years.

After Madsen left Alaska in the weeks following the earthquake, it took her a while to stop her wandering. Her father and stepmother were overjoyed to have her home, safe and sound, in Long Beach, but they thought she was crazy to have brought Norman Selanoff with her. The tension was palpable, so he left after a few days, eventually ending up back in Alaska.

After a month or so at home, Madsen was getting restless again. Her aunt Muriel and her husband were planning a long summer trip to Europe with their daughters and a cousin. Maybe it would be good for Kris to join them, to get her mind off Alaska and what she'd been through. The plan was to buy a Volkswagen in Germany and explore the continent. It would be a simple thing, they said, to buy another plane ticket, and Kris could share hotel rooms with the cousins. Madsen needed to think only briefly before agreeing to join them. She had plenty of money from her time in Alaska, and Muriel's family would be good traveling companions. She told her stepmother that perhaps she'd find a job in Europe and stay.

They left in June. Less than a week later Madsen was in a hotel room in Rome, idly looking at the classified ads in the *International Herald Tribune.* One of them caught her attention. It was from the International School in Frankfurt, West Germany, seeking an elementary teacher for the fall. She called, and a few days later, with her accommodating relatives having agreed to a short detour from their planned driving route, was in Frankfurt for an interview. In August, when they were in England, she found out she'd gotten the job. She didn't return to Long Beach till the following summer.

At the Frankfurt school she met another American teacher, Drew Van Winkle. The two hit it off immediately, though it took a while, and some transatlantic back-and-forth, for romance to blossom. They were married in 1966, Kris gave birth to twins in 1968 and the following year they returned to the United States to pursue graduate degrees and be closer to their parents. They settled in the Gold Country of Northern California and taught in the public schools for decades before retiring.

Kris Van Winkle has never been back to Alaska. But Chenega, and what happened there in 1964, has never been far from her mind. Yet she was only vaguely aware of the villagers' story since then.

Most of the Chenegans were never completely comfortable in Tatitlek, where the government had moved them after the quake, and some started to drift away that first fall. Others followed suit later. Within a few years most were scattered around Alaska, even Outside in the Lower 48. Avis Kompkoff, for one, eventually ended up back in Cordova, where she lives to this day. Nick Kompkoff, who had become a lay priest, and others tried to maintain a kind of community in exile, with village meetings at least once a year, attended by whoever could get to them. In the back of everyone's mind was the idea of rebuilding Chenega. In the immediate aftermath of the quake the villagers had resolved never to return to the island, but as the years passed that resolve began to weaken. Perhaps Chenega village would rise again someday.

In 1968, another event occurred that would change Alaska forever. This time it wasn't an earthquake; rather, it was the discovery of oil on the North Slope of the Brooks Range, in the far north. The oil itself would have a great impact, of course, but what affected life in Alaska much more immediately was the need to do something about natives' claims to much of the state's territory. The issue had been shelved during the statehood debate ten years before, but now, if Alaska was to be able to fully exploit what appeared to be vast oil reserves by building a pipeline across the tundra to bring the oil to market, the claims had to be resolved.

The solution was the Alaska Native Claims Settlement Act, passed by Congress in 1971. In exchange for relinquishing their claims to the bulk of the state's land (and their rights to fish and hunt where they pleased), the native population received more than forty million acres outright. They were also allowed to establish regional and village for-profit corporations.

The Chenega Corporation was created several years later, with sixty-eight shareholders. As its share of the settlement, it received about seventy-five thousand acres of land in Prince William Sound. That helped fuel a desire to reestablish the village at Chenega Island. But a visit to the site by many villagers in 1976 quickly quenched that idea. Even more than a decade later, the memories of the quake and of lost loved ones were too strong.

Kompkoff and others, notably Larry Evanoff, who had been off at school when the quake struck but had lost his parents in the disaster, kept pursuing the goal of bringing the Chenegans together physically. They found some land at Crab Bay on Evans

Island, about fifteen miles due south of the old village site by water. It was about seventy feet above the water and had not been inundated during the 1964 quake. There was plenty of space for homes, businesses and a church, even an airport.

The Chenega Corporation and the village council voted in 1977 to establish the new village of Chenega Bay at the Evans Island site. Money for houses, a school, docks and other improvements was a problem, but over the years enough was raised so that the bare bones of a village began to take shape. The first Chenegans moved there in 1982, and others followed beginning in 1984.

Five years later, yet another disaster struck Alaska, this time made by man. The *Exxon Valdez,* a supertanker carrying some of the millions of barrels of North Slope crude oil that had begun flowing in the 1970s, ran aground in Prince William Sound, not far, in fact, from the village of Tatitlek. More than ten million gallons of heavy oil spread across the sound, fouling the rocks, beaches and fishing grounds. Chenegans, some of whom had returned to a subsistence lifestyle based on hunting and fishing, were badly affected as fish, seals and other sea creatures were contaminated.

The oil spill and the lengthy cleanup that followed made establishing a new community in Chenega Bay even more difficult. Today about eighty people are living there. But not all are natives, and some Chenegan natives live in Anchorage, tending to the affairs of the Chenega Corporation, which has developed as a business, specializing in government contracts that it, as a native corporation, can bid on favorably.

Back in old Chenega, the villagers have built a pavilion for the ceremony they conduct once a year, when they go back in the spring to remember those who died. The schoolhouse on the hill still stands, but only barely—the whole structure is in a state of slow collapse. What was once a bare knoll is now overgrown with spruces. The clearing where the church and houses stood is dense with alders. On the beach, a few of the pilings that once held up the dock are still visible, and the short stretch of the bulkhead that remains is rotting and falling apart.

With the fiftieth anniversary of the quake in 2014, Kris Van Winkle decided she needed to do something. During her stay in the village, she'd taken some photographs from up on the hill. They showed the cove and the dock, the church and a few of the houses. Most of her photos, however, were of her students, taken at the school. One or two of them showed the children performing at the "American" Christmas show she had organized for parents. But most of the photos showed the students—including the two she'd lost, Julia Kompkoff and Cindy Jackson—outside, smiling and laughing in the sun. She'd even taken a few photos up on the hill after the quake, when the survivors gathered for the night. You could see the fear and sadness on some of their faces.

Van Winkle put together a portfolio of 8×10 prints of some of the photos and wrote a note to the people of Chenega, telling them she hadn't forgotten what they'd gone through that day. She put it all in an envelope and sent it care of the Chenega Corporation in Anchorage. She never heard back.

She and Drew showed the photos to me, displaying them on a television in their living room. As Van Winkle clicked through photo after photo, she talked about life in the village and about the events of that terrible day. The thing that had stuck with her, she said, was the power of the earthquake—she'd never forgotten how the water in the cove had disappeared—and the stoicism of the survivors, who had lost everything but still somehow had the strength to go on.

Tom McAlister steered his pickup truck down Hazelet Avenue, past Iditarod Street toward Egan Drive, which counts, more than any other, as Valdez's main street. It was late winter 2015, and McAlister, who had lived and worked here since before the quake—he was the Valdez fire chief for years, among other jobs—was showing me around "new" Valdez, which by now was nearly half a century old. He was pointing out some of the more than fifty homes and other buildings that had been moved from the old town, four miles away.

Over there, behind some trees, was the house that Owen Meals had lived in. Over here was a house that was moved only because the couple that owned it had bought it for $6,000 shortly before the earthquake. They hadn't wanted to keep it but couldn't get out of the mortgage, so they moved it instead.

McAlister had first come to Alaska just after graduating from high school in Snohomish, Washington. He'd gotten on a boat bound for Valdez, found a job and stayed the summer. He'd come back for good two years later, in 1959. There he met and married Gloria Dewing, whose family had come to Valdez a few years before. They were home in their house on McKinley Street with their first child, a boy who was just six weeks old, when the quake hit. They'd made it through the disaster, and while Gloria took the infant and went Outside for a while, Tom had stayed and helped with the recovery. Later, when the family was reunited, they had moved into a trailer, as their house on McKinley was uninhabitable. The trailer was moved over to the new town; they lived in it for another six years until they bought a house. "We learned about wind," McAlister said. Old Valdez, with its mature willows and spruces and other trees, had been largely protected from the winds that came rushing off the water or down from the glacier. But the new town was practically bare. Another trailer on their block rolled over in one particularly fierce windstorm that first winter. But the McAlisters learned to cope.

By 1967, there was no one left in the old town. New Valdez was full of people—the population was about what it had been prequake—but it was struggling. Even before the move was completed there were efforts to increase tourism, as that seemed to be one of the few activities that held any promise for improving the town's economic prospects, since what little shipping was left had been effectively killed by the earthquake. There was even a proposal to make the old town site into a tourist attraction, a historic park that would preserve some of the buildings and focus on the town's role in the Klondike gold rush and, of course, the earthquake. Plans were drawn up, but the idea didn't get very far. Cost was an issue, but there was a more fundamental problem: the land was still unstable. How could anyone in good conscience invite tourists to a place that might at any moment be shaken apart? With those plans dead, old Valdez was left to rot—and occasionally burn, as it became a prime target for arsonists—until the town voted to raze the remaining buildings and clear the land.

Prospects for the new Valdez weren't all that much better. In 1968, however, the same event that changed the lives of native Alaskans changed Valdez forever. With the discovery of oil on the North Slope came a crucial question: how would it be delivered to an oil-hungry world? In the summer of 1969, with much ballyhoo, Humble (now ExxonMobil) sent an icebreaker-tanker, the SS *Manhattan,* through the Northwest Passage across the Canadian Arctic. But a later attempt at a second transit in winter, with more and heavier ice, proved impossible. That, and environmental concerns about the risk of a spill to the pristine Arctic, killed the tanker idea, but there were others, including a scheme to extend the Alaska Railroad north from Fairbanks and ship the oil by tank car, with Anchorage or Seward becoming an oil port. But the logistics of such an operation were daunting at best.

The best possibility, and one eventually favored by the oil companies, seemed to be a pipeline. But where to put it? A proposal to run it down to Fairbanks and then along the railroad right-of-way had many backers. But Valdez officials had their own ideas. Their town was closer than Seward, and it had a deepwater port that was, as they would proudly tell anyone, ice-free all winter long. As for earthquake concerns, there was enough stable land for a terminal and operations center on the southern side of the bay, across the Lowe River, where Fort Liscum had been built at the turn of the century.

The oil companies sent in their own geologists and soon were developing a proposal to run the pipeline along the Richardson Highway, over Thompson Pass, to Valdez. Environmental organizations sued, holding up construction for more than three years, but a federal law approved in 1973 allowed the project to proceed.

There had been some funny business: pipe for the pipeline had begun showing up in Valdez long before construction was given the go-ahead. But the town was happy to have the project. Thousands of workers came to Valdez, where the population swelled to more than eight thousand. The now-cleared old town site turned out to be the perfect place to stage pipe and other materials for the construction work.

It was almost too much of a boom. There were so many construction workers in Valdez, some of dubious character, that residents felt that in some ways their town had been taken away from them. They stayed away from the bars, restaurants and other places that were now inundated with workers from elsewhere. There were a lot of house parties those years, one old-timer recalled.

The construction crush eventually ended, and oil first flowed from the pipeline in June 1977. The town shrank somewhat, but not completely, as people stayed on to work at the pipeline control center and at the terminal where the supertankers were loaded. The *Exxon Valdez* spill in 1989 brought more people back as Valdez became the center of cleanup operations.

The pipeline jobs changed the town. In the past—certainly before the earthquake—there hadn't been many full-time, year-round jobs. People scraped by, working for the highway department in the winter and building homes in the summer, fishing part of the year and longshoring when they could—the men who perished at the dock during the earthquake, after all, had been part-timers picking up extra cash. That same year, Tom McAlister recalled, when it came time to do his taxes he'd had seven W-2 forms from the various jobs he'd held.

But the pipeline changed that, with steady jobs that paid good wages. Valdez was less of a catch-as-catch-can sort of place, and neighbors were less reliant on one another than they once had been.

Of course, the oil will run out someday, probably sooner rather than later, and those jobs will disappear. The town is already on something of a downward trajectory, as flow through the pipeline has steadily declined from its peak of about two million barrels a day in the late 1980s to about one-quarter of that now. The writing is on the wall, and town officials are trying to boost tourism again.

That's the nature of Valdez, though, McAlister said as we continued the house tour. It's always been boom or bust. He'd lived through all of it and still loves the town, although he acknowledged that it wasn't the same place it had been in the days before the quake.

We'd reached Egan Drive, named for the town's favorite-son governor, and what passed for downtown Valdez. The vast parking lots that residents had insisted were needed as a place to dump snow stretched out before us. Old Valdez hadn't had much of a downtown either, but it was compact, with cheek-by-jowl businesses like the Pinzon Bar and Gilson's grocery and Woodford's dry goods. New Valdez was tidier, but it didn't have as much character. "That's one of the things we lost," McAlister said.

A few days before we visited Alaganik Slough, Plafker and I headed out into Prince William Sound with Peter Haeussler, a USGS geologist based in Anchorage. Haeussler has spent much of his career studying the 1964 earthquake—among other findings, he and colleagues identified the likely source of the undersea sediments that collapsed and caused the waves at Chenega. A great admirer of Plafker, he arranged for a boat, and we left from Whittier, bound for Montague Island. We anchored in a spot on the island's southwestern side that Plafker had named Fault Cove because one of the splay faults passes through it. We ran our skiff onto shore and started hiking through a tangle of alders toward higher ground in the distance.

These splay faults are still a subject of great interest to geologists. Haeussler and some of his colleagues had recently published a paper about them, showing how, in addition to their sudden movement during the 1964 quake, they were slowly uplifting over time. It was now known that there were more than two of them, but where we were hiking was between the two that Plafker had identified in the summer of 1964—the scarp formed by the Hanning Bay Fault when it lifted up was immediately to our left, and the Patton Bay Fault was perhaps four miles to our right.

The high ground we were headed for was, in fact, the back side of the Patton Bay Fault, where some of the highest uplift from the quake was measured. As we turned uphill, the alders gradually gave way to spruces and an understory thick with vines and the pernicious shrub called devil's club. The two geologists had warned me about this: the plant's thick stems tempt a hiker who is looking for something to grab on to to help scramble up a slope, but they're covered with tiny spines that will cause a rash and take forever to remove from the skin.

Soon we emerged from the spruces, whacked our way through another narrow thicket of alders and found ourselves on a less overgrown slope covered in rocks and carpeted with moss. I bent down and examined the rocks. They were cobbles, made smooth, obviously, by the action of water. This, I realized and Plafker and Haeussler quickly confirmed, was a cobble beach. Rather, it *had been* a cobble beach, until 5:36 p.m. on March 27, 1964. Then, in the course of a few minutes, it had risen up more than thirty feet, and the sea had bothered it no more. I'd seen a lot of signs of the earthquake—the old sites in Valdez and Chenega, Portage and its drowned trees, Earthquake Park in the Turnagain neighborhood in Anchorage where the tortured land is still visible—but this was the most impressive evidence yet of the forces that were unleashed on that day.

A few days later, in Cordova, Plafker showed me another impressive sight. We had driven out of town on Route 10, but in the opposite direction from Alaganik Slough. Hugging the shoreline along Orca Inlet, the road here, too, leads to nowhere: it ends about four miles out of town at the site of an old cannery. There's an adventure lodge there now, offering fishing, kayaking and other activities to tourists. But some of the old cannery buildings remain. They are in various states of disrepair, as are the docks where the fishing boats once tied up to unload their catch. This area was uplifted about six feet during the quake, so the docks seem oddly high in relation to the water. High tide barely reaches some of the wooden pilings; before the quake they would have been well underwater.

We walked over and looked at one of the pilings more closely. Plafker showed me a ring of living barnacles around the wood near his feet. Then he reached up and pointed to a spot above his head. Though it was not quite as obvious, there was another ring there, the white remains of long-dead organisms still encrusted on the piling. Age and exposure had eliminated all traces of it most everywhere else, but here the barnacle line, which Plafker had so relied upon to help understand the greatest earthquake North America has ever known, survived.

At Alaganik Slough the next day, our surveying work over, Plafker and I wandered up and down the main channel, with me struggling to keep up. We stopped from time to time to look at cuts in the banks that showed alternating layers of peat and mud. These, Plafker said, were signs of ancient earthquakes. Having studied the Copper River delta for years, he knew the earthquake sequence here well. A quake occurs and land that had been below the water line is lifted up. Vegetation, including mosses, begins to grow and a thick layer of peat starts to develop. But the land slowly sinks over time, in part because of the weight of all the glacial sediment being brought down by the river. Eventually it has sunk so much that some of the peatland is low enough that the tidal water rushes over it again, depositing sediment that compresses the peat. Eventually another earthquake occurs, the land is uplifted once again and the process repeats itself.

The alternating layers in the channel banks represent different earthquakes, and dating the peat gives an idea of when each happened. That's what Plafker's cores were all about. From earlier work he already had a good idea of how often large megathrust earthquakes had happened in Alaska. But you could never have enough data.

Plafker stopped to talk about the studies he'd done in the

delta. It was similar in some ways to research that had been done in the Pacific Northwest, he said. There, they knew that major quakes could happen as little as a few hundred years apart, and that the last one was more than three hundred years ago. That's why there was so much concern. Alaskans, on the other hand, could probably breathe easy. Major megathrust quakes happened here about every six hundred to eight hundred years, so another big one was probably centuries off. That was what his research indicated, anyway. Plafker turned to head back up the muddy channel. He wished he could be around when the next one happens, he said, to find out if his estimates were correct.

ACKNOWLEDGMENTS

When writing nonfiction, a cooperative subject can mean the world. In the case of George Plafker, "cooperative" is a wholly inadequate word. George spent hours patiently telling me his life story and explaining the ins and outs of geology and plate tectonics, answering every one of my dumb questions. But beyond that, George and his partner, Doris Coonrad, graciously welcomed me into their California home many times. Thanks for the hospitality, including the countless delicious meals prepared by Doris, caterer extraordinaire.

Likewise, Kris and Drew Van Winkle warmly welcomed me when I visited them. I owe much gratitude to Kris for sharing her story and amazing photographs of her time in Chenega, and to both of them for the lighthouse tour. Special thanks to Mads Tudvad Jensen for helping me find Kris.

Peter Haeussler of the Geological Survey in Alaska has for years been pointing out to anyone who will listen the work that George did and the role it played in the acceptance of plate tectonics. Peter's writings for the Survey around the time of the fiftieth anniversary of the quake were what got me interested in the subject in the first place. Thanks to Peter for all that, and for taking George and me out to some of the places in Prince William Sound that George studied a half century before. And thanks to Marti L.

Miller, chief of the geology office at the Survey's Alaska Science Center in Anchorage, and her staff for the use of the boat.

More than half a century later, the Alaska earthquake is still a source of pain for many who lived through it. I've respected the wishes of those who don't want to talk about what they experienced. But I'm grateful to the people who did, as well as other people who provided invaluable insight or assistance: Arthur Grantz, Peter Molnar, Dan Kendall, Gary Minish, Dorothy Moore, Gloria Day, Gloria and Tom McAlister, Avis Kompkoff, Ross Stein, Rita Miraglia, John F. C. Johnson, Nancy Yaw Davis, Steve Ranney, Suzanne Cook Taylor, Kevin Krajick and the many USGS media relations people who have helped me over the years. Thanks also to Andrew Goldstein, curator of collections, and the staff of the Valdez Museum and Historical Archive, as well as the staffs of the Alaska and Polar Regions Collections and Archives at the Elmer E. Rasmuson Library, University of Alaska–Fairbanks; Prince William Sound College library; the Archives and Special Collections at the University of Alaska–Anchorage library; Alaska Resources Library and Information Services; the Alaska Collection at the Z. J. Loussac Public Library, Anchorage; the Anchorage Museum; Alaska's Digital Archives and the Alaska State Library Historical Collections.

I'm grateful to my editors at the *New York Times* for allowing me to step back from daily journalism for a while. My editors in the science department, including the late Barbara Strauch, Celia Dugger, Mary Ann Giordano and Adam Bryant, offered encouragement and support, as did my reporting podmates. Thanks for shelter, food and moral support to Harriet Moss and Ernesto Sanchez, and to Steve and Ted Ballou, their families and Troy Crisswell for the musical interludes.

This project began when an editor at a New York publishing house read my article in the *Times* about the fiftieth anniversary of the Alaska quake. That editor was Roger Scholl, at Crown, and he thought the subject of the quake and George Plafker's work might make for an interesting book. Many thanks to him for that inspiration, and for helping guide and shape the book as it progressed. Thanks also to others at Crown, including Julia Elliott and Craig Adams.

My agent, Gillian MacKenzie, has for a long time encouraged me to pursue book writing. For that, and for her invaluable advice and ideas about this book, I'm grateful.

Lastly, thanks to Savannah and Walker for putting up with all of this, and for keeping the dogs at bay.

NOTES AND SUGGESTIONS FOR FURTHER READING

1. Altered State

The account of the geologists' first few days in Alaska following the earthquake, and the conditions they found, is based largely on interviews with George Plafker and Arthur Grantz and on the US Geological Survey's professional papers on the earthquake (see Additional Sources).

6 **The bases had not been too badly damaged:** Some of the damage at Elmendorf Air Force Base is described in National Research Council, *The Great Alaska Earthquake of 1964,* vol. 6, *Engineering* (Washington, DC: National Academy of Sciences, 1973).

7 **military had spent billions:** For an interesting look at the effect of the Cold War military buildup on Alaska, see Laurel J. Hummel, "The U.S. Military as Geographical Agent: The Case of Cold War Alaska," *Geographical Review* 95, no. 1 (January 2005): 47–72.

9 **"one of the richest, most glorious mountain landscapes":** John Muir, in Edward H. Harriman, *Harriman Alaska Expedition* (Washington, DC: National Academy of Sciences, 1910), 1:132.

2. Under the Mountain

The description of Chenega Island is compiled from various sources, including Kaj Birket-Smith, *The Chugach Eskimo* (Copenhagen: National Museum Publications Fund, 1953); Donald R. Poling, comp., *Chenega Diaries: Stories and*

Voices of Our Past (n.p.: Chenega Corporation, 2011); John Smelcer, *The Day That Cries Forever* (Anchorage: Todd Communications, 2006); William E. Simeone and Rita Miraglia, *An Ethnography of Chenega Bay and Tatitlek, Alaska* (Anchorage: Alaska Department of Fish and Game, 2000); and interviews with Kris Van Winkle and Avis Kompkoff.

19 **"They were unwilling, however, to venture along-side":** This and other quotations and descriptions of Cook's time in Prince William Sound are from Captain James Cook, *Captain Cook's Third and Last Voyage to the Pacific Ocean* (London: Fielding and Stockdale, 1785).

20 **They viewed the Russian interlopers as aliens:** Birket-Smith, *Chugach Eskimo,* 10.

20 **One Russian captain wrote:** P. A. Tikhmenev, *A History of the Russian-American Company* (Seattle: University of Washington Press, 1978), 45.

20 **The Chugach themselves divided the sound:** Birket-Smith, *Chugach Eskimo,* 20.

21 **"The whole existence of the Chugach":** For descriptions of the subsistence lifestyle, see Simeone and Miraglia, *Ethnography of Chenega Bay;* and Birket-Smith, *Chugach Eskimo.*

21 **"soaked in grease":** Birket-Smith, *Chugach Eskimo,* 21.

24 **"the kingdom of death":** Ieromonk Serafin, "The Kingdom of Death," *American Orthodox Messenger,* 1907, cited in Simeone and Miraglia, *Ethnography of Chenega Bay.*

25 **Another lasting Russian influence:** An overview of the Russian Orthodox Church in Alaska is in Claus M. Naske and Herman E. Slotnick, *Alaska: A History,* 3rd ed. (Norman: University of Oklahoma Press, 2014).

26 **"If you're not good":** Interview with Kris Van Winkle.

27 **an especially joyful, even raucous, time:** The holiday is discussed in detail in Poling, *Chenega Diaries.*

28 **He'd gotten the name:** Ibid., 274.

3. An Accident of Geography

29 **one that he would not soon forget:** Abercrombie described his experiences in a number of reports, which are included in W. R. Abercrombie et al., *Compilation of Narratives of Explorations in Alaska* (Washington, DC: Government Printing Office, 1900), and which form the basis for the description of conditions in Valdez during the gold rush.

31 **"A more motley-looking crowd":** Ibid., 758.

32 **"During my 22 years of service":** Ibid., 568.

33 **the Canadian government imposed rules:** For an excellent overview of the gold rush, see Pierre Berton, *The Klondike Fever: The Life and Death of the Last Great Gold Rush* (1958; repr., New York: Basic Books, 2003).

34 **Ships could beach themselves:** This is described in, among other sources, Abercrombie et al., *Compilation of Narratives.*

35 **A glacier such as the one:** David Evans and Douglas Benn, *A Practical Guide to the Study of Glacial Sediments* (London: Arnold, 2004).

36 **This was recognized as a problem:** Abercrombie et al., *Compilation of Narratives,* 808.

37 **Valdez might remain a mining hub:** It was Cordova, however, that ultimately benefited from the copper ore in the Wrangells. The story is told in Janson, Lone E., *The Copper Spike* (Anchorage: Alaska Northwest Publishing, 1995).

38 **Among the thousands of adventurers:** The story of Hazelet and Cheever, including the account of eating a marmot, is told by Hazelet in John H. Clark, ed., *Hazelet's Journal* (Louisville, KY: Old Stone Press, 2012).

39 **Some of the operations were fabulously successful:** George Sundborg, *Valdez Industrial Report* (Juneau: Alaska Development Board, 1955).

40 **One of Valdez's selling points:** Ibid.

41 **High school students often brought their rifles:** Interviews with Dan Kendall and Gary Minish.

41 **billing itself as the "Switzerland of Alaska":** The slogan was found on everything from restaurant place mats to tourism brochures to a wooden

sign in the center of town. Examples can be found in the collection of the Valdez Museum and Historical Archive, www.valdezmuseum.org.

43 **A local Democrat, complaining once:** Interview with Gloria Day.

4. Clam Broth and Beer

47 **known to everyone who lived there as the I I:** The description of life at the orphanage is from interviews with George Plafker and from Ira A. Greenberg, ed., *The Hebrew National Orphan Home: Memories of Orphanage Life* (Westport, CT: Bergin and Garvey, 2001).

49 **a gentlemanly New Jerseyan:** George T. Faust, "Memorial of Alfred Cary Hawkins," *American Mineralogist* 54 (1969): 619–25.

52 **the corps had been kept busy:** For a history of the corps' Sacramento District, see Willie R. Collins et al., *Sacramento District History (1929–2004)* (Sacramento, CA: US Army Corps of Engineers, 2005).

53 **The USGS was established in 1879:** For an overview of the history of the Survey, see Mary C. Rabbitt, *The United States Geological Survey: 1879–1989* (Washington, DC: Government Printing Office, 1989).

5. The Floating World

The development of the theory of plate tectonics is exceptionally well told, by the scientists involved, in Naomi Oreskes, ed., *Plate Tectonics: An Insider's History of the Modern Theory of the Earth* (Boulder, CO: Westview Press, 2001).

56 **more than 130 have erupted:** Up-to-the-minute information about Alaska's volcanoes can be found at www.avo.alaska.edu.

56 **expedition to explore the devastation:** Robert Fiske Griggs, *The Valley of Ten Thousand Smokes* (Washington, DC: National Geographic Society, 1922).

57 **when he was a distinguished scientist:** Hans Cloos, *Conversation with the Earth* (New York: Alfred A. Knopf, 1953).

58 **The first spark of the idea:** The most recent, most complete, and perhaps best biography of Alfred Wegener is Mott T. Greene, *Alfred Wegener:*

Science, Exploration and the Theory of Continental Drift (Baltimore: Johns Hopkins University Press, 2015).

60 **"loosened the continents from the terrestrial core":** Cloos, *Conversation with the Earth,* 396.

63 **The gist of his theory:** Alfred Wegener, *The Origin of Continents and Oceans* (1966; repr., New York: Dover Publications, 2011).

64 **This sentiment was particularly felt:** For a fuller account of the New York meeting, see Homer E. LeGrand, *Drifting Continents and Shifting Theories* (Cambridge: Cambridge University Press, 1989).

65 **Harry Hess laid the groundwork:** Hess's life is concisely described in Harold L. James, *Harry Hammond Hess, 1906–1969* (Washington, DC: National Academy of Sciences, 1973).

67 **another scientist had speculated:** K.C. Dunham, "Arthur Holmes: 1890–1965," *Biographical Memoirs of the Fellows of the Royal Society* 12 (1966): 290–310.

67 **he estimated that the planet was 1.6 billion years old:** Arthur Holmes, *The Age of the Earth* (London: Harper, 1913). Holmes was off by about three billion years.

68 **Holmes suggested in a paper:** Arthur Holmes, "Radioactivity and Earth Movements," *Transactions of the Geological Society of Glasgow* 18 (1929): 559–606.

68 **he described his idea as speculative:** Arthur Holmes, *Principles of Physical Geology* (London: Thomas Nelson and Sons, 1944).

70 **he proposed in 1959:** Harry Hess, "Nature of the Great Oceanic Ridges," International Ocean Congress preprints, American Association for the Advancement of Science, Washington, DC, 1959, 33–34.

70 **as Hess himself acknowledged:** Harry Hess, "A History of Ocean Basins," in *Petrologic Studies: A Volume to Honor A. F. Buddington* (New York: Geological Society of America, 1962), 599–620.

6. *Spiking Out*

73 **You examine the rock formation:** This and other descriptions of geology fieldwork are from interviews with George Plafker and from Dougal Dixon and Raymond L. Bernor, eds., *The Practical Geologist* (New York: Simon & Schuster, 1992).

74 **He'd joined the Alaska branch in 1942:** "Memorial: Don John Miller," *Bulletin of the American Association of Petroleum Geologists* 46, no. 8 (August 1962): 1534–37.

84 **Miller and MacColl had been working:** An account of the accident is in the *San Mateo Times and Daily News Leader* (CA), August 8, 1961.

7. *Before the Storm*

87 **the state had decided to allow:** The plan to cull wolves was eventually dropped. Timothy Egan, "Facing Boycott, Alaska Drops Plan to Kill Wolves," *New York Times*, December 23, 1992.

88 **"We can't just let nature run wild":** Malcom B. Roberts, ed., *The Wit and Wisdom of Wally Hickel* (Anchorage: Searchers Press, 1994).

89 **The valley had seen its own development:** For a history of the Matanuska Colony, see Helen Hegener, *The 1935 Matanuska Colony Project: The Remarkable History of a New Deal Experiment in Alaska* (Wasilla, AK: Northern Light Media, 2014).

90 **no fewer than thirty-three saloons:** Harry Ritter, *Alaska's History: The People, Land and Events of the North Country* (Anchorage: Alaska Northwest Books, 2015).

92 **Alaska's people were "enterprising, vigorous, warmhearted, modern":** Ernest Gruening, "Alaska Proudly Joins the Union," *National Geographic*, July 1959.

95 **Down in the village:** The description of events in Chenega on the day of the earthquake is derived from various accounts, including interviews with Kris Van Winkle and Avis Kompkoff; John Smelcer, *The Day That Cries Forever* (Anchorage: Todd Communications, 2006); recollections of elders in Donald R. Poling, comp., *Chenega Diaries: Stories and*

Voices of Our Past (n.p.: Chenega Corporation, 2011); unpublished survivor interviews by Alaska Department of Fish and Game employees; and newspaper articles.

96 **he wandered over to the Smokehouse:** *Cordova Times,* March 29, 1975.

96 **Kenny had wanted to chase and throw stones:** Smelcer, *Day That Cries Forever.*

97 **The two had gone out in a skiff:** Interview with Avis Kompkoff.

99 **the people of Valdez were suckers:** Descriptions of Valdez at the time of the earthquake are based on interviews with Tom and Gloria McAlister, Gary Minish, Dan Kendall, Gloria Day and Dorothy Moore; Karen LaChance, *Valdez: A Brief Oral History* (Valdez, AK: Prince William Sound Community College, 1995); and newspaper articles.

100 **For a large ship, about a dozen men were needed:** George Sundborg, *Valdez Industrial Report* (Juneau: Alaska Development Board, 1955).

101 **that steel had failed spectacularly:** For a history of the Liberty ship program, see Peter Elphick, *Liberty: The Ships That Won the War* (Annapolis, MD: Naval Institute Press, 2006). An account of the near-sinking of the *Chief Washakie* appears in *Time* magazine, March 20, 1944.

103 **packed themselves into the family car:** Interview with Gary Minish.

104 **She and Walter had spent much of Thursday:** Interview with Gloria Day.

8. *Faults*

107 **Howard Ulrich maneuvered:** The near-sinking of the *Edrie* is told in the *Daily Sitka Sentinel,* February 25, 1988, and July 9, 1996.

109 **The next day, Don Miller:** Don J. Miller, *Giant Waves in Lituya Bay, Alaska,* USGS Professional Paper 354-C (Washington, DC: US Geological Survey, 1960).

109 **In ancient times, the shaking of the earth:** A concise history of the rise of seismology is found in Benjamin F. Howell Jr., *An Introduction to Seismological Research: History and Development* (Cambridge: Cambridge University Press, 1990).

111 **Gilbert was born in upstate New York:** William M. Davis, *Grove K. Gilbert, 1843–1918* (Washington, DC: National Academy of Sciences, 1922).

112 **a lesser-known report:** Grove K. Gilbert, "A Theory of Earthquakes of the Great Basin, with a Practical Application," *Journal of American Science*, 3rd ser., 27, no. 57 (January 1884): 49–53.

113 **seismometers had been around in one form or another:** Howell, *Introduction to Seismological Research*, 58–72.

116 **The shocks also shattered Muir Glacier:** Ralph S. Tarr and Lawrence Martin, *The Earthquakes at Yakutat Bay, Alaska, in September 1899* (Washington, DC: US Geological Survey, 1912).

9. Shaken

Sources of personal accounts of the earthquake include Genie Chance, *Chronology of Physical Events of the Alaskan Earthquake* (n.p.: National Science Foundation, 1966); various USGS professional papers on the quake; John Smelcer, *The Day That Cries Forever* (Anchorage: Todd Communications, 2006); Karen LaChance, *Valdez: A Brief Oral History* (Valdez, AK: Prince William Sound Community College, 1995); George Plafker's unpublished field notebooks; and interviews with survivors.

117 **an episode of the sci-fi marionette series:** Many personal accounts of the earthquake by children in Anchorage mention watching this show.

117 **At her medical office south of downtown:** Chance, *Chronology of Physical Events*, 101. This is the source of the accounts of Dean Smith, Tobias Shugak and others as well.

119 **convinced Russian battleships were shelling:** From Plafker's unpublished field notebooks.

119 ***his* car wasn't the problem:** William P. E. Graves, "Earthquake!," *National Geographic*, July 1964.

120 **The mayor of Anchorage, George Sharrock:** Unpublished interview with Mayor George Sharrock by Glenn Bordwell, Anchorage, in University of Alaska–Fairbanks' Alaska and Polar Regions Collections.

121 **the postmaster reported afterward:** Chance, *Chronology of Physical Events,* 94.

122 **She started skiing toward her colleagues:** There are several accounts of what happened at the lake, including ibid., 95.

122 **Blanche Clark had just left:** There are many accounts of Clark's experience outside the J. C. Penney store, including ibid., 133.

122 **Carol Tucker was on the third floor:** Graves, "Earthquake!"

124 **Atwood had just gotten home:** Atwood wrote about his ordeal in his newspaper, the *Anchorage Daily Times,* on March 29. Chance, *Chronology of Physical Events,* has another account based on an interview.

127 **She wasn't unfamiliar with earthquakes, either:** Interview with Avis Kompkoff.

128 **Nick Kompkoff's first thought:** *Cordova Times,* March 29, 1975.

128 **Timmy Selanoff watched in amazement:** Smelcer, *Day That Cries Forever,* 54.

129 **Recollections of survivors in the days that followed:** Interviews conducted by Plafker, in his unpublished field notebooks.

131 **He remembered grabbing on to a twig:** Smelcer, *Day That Cries Forever,* 55.

135 **The screeching was the worst thing:** Interview with Gloria Day.

135 **He made it up to the bridge:** There are several accounts of what happened to the *Chena,* by Captain Stewart and others, in National Research Council, *The Great Alaska Earthquake of 1964,* vol. 6, *Engineering* (Washington, DC: National Academy of Sciences, 1973); Henry W. Coulter and Ralph R. Migliaccio, *Effects of the Earthquake of March 27, 1964, at Valdez, Alaska,* USGS Professional Paper 542-C (Washington, DC: US Geological Survey, 1966); "The Bouncing Chena," *Alaska Construction,* May–June 1965.

136 **a couple of shutterbugs:** Footage that these men shot can be seen in a short video, *1964 Good Friday Earthquake, Valdez* (University of Alaska–Fairbanks, Bartlett Collection, AAF-1438). An analysis of frames from the films is in National Research Council, *The Great Alaska Earthquake of 1964,* vol. 5, *Oceanography and Coastal Engineering* (Washington, DC: National Academy of Sciences, 1972).

136 **The people on the dock who had been working:** The accounts of Dorney and others are in the *Seattle Daily Times,* April 4, 1964.

138 **His son was standing:** Described in Coulter and Migliaccio, *Effects of the Earthquake.*

138 **Gilson and his customers experienced this:** LaChance, *Valdez.*

10. Stunned

143 **one quick calculation:** The most extreme estimate, by the chief of seismology at the US Coast and Geodetic Survey, was that the quake released ten million times more energy than the bomb that was dropped on Hiroshima. It was reported in many newspapers around the country.

144 **a barge loaded with lumber:** Interview with George Plafker.

144 **It eventually spread out:** Austin Post, *Effects of the March 1964 Alaska Earthquake on Glaciers,* USGS Professional Paper 544-D (Washington, DC: Government Printing Office, 1967).

144 **Estimates of the duration:** Genie Chance, *Chronology of Physical Events of the Alaskan Earthquake* (n.p.: National Science Foundation, 1966).

148 **a characteristic of the missiles:** For a short history of Nike missiles in Alaska, see Kristy Hollinger, *Nike Hercules Operations in Alaska: 1959–1979* (Fort Richardson, AK: US Army Alaska, 2004), www.a-2-562.org/.

151 **left a trail of destruction:** For details on the waves that hit the Pacific Northwest, see National Research Council, *The Great Alaska Earthquake of 1964,* vol. 5, *Oceanography and Coastal Engineering* (Washington, DC: National Academy of Sciences, 1972).

153 **The couple grabbed the children:** An Associated Press account of the events on the beach was published widely, including in the *Fairbanks Daily News–Miner,* April 3, 1964.

153 **a small port in timber country:** The destruction in Crescent City is described in detail in Wallace H. Griffin, *Crescent City's Dark Disaster* (Crescent City, OR: Crescent City Press, 1984).

160 **Ashen received a call:** Account from a 1964 KPIX-TV (San Francisco) documentary about the earthquake, *Chenega Is Gone.* Ashen's work in

Chenega is also described in Jack Foisie, "Rebirth of a Village," *San Francisco Chronicle*, ca. May 30, 1964.

163 **On Knight Island they found the second floor:** Interview with Avis Kompkoff.

164 **never had a great affinity:** William E. Simeone and Rita Miraglia, *An Ethnography of Chenega Bay and Tatitlek, Alaska* (Anchorage: Alaska Department of Fish and Game, 2000).

165 **Ashen told a newspaper reporter later:** *San Francisco Chronicle*, June 3, 1964.

165 **The US Army showed up:** Sources for the description of Valdez after the earthquake include National Research Council, *The Great Alaska Earthquake of 1964*, vol. 7, *Human Ecology* (Washington, DC: National Academy of Sciences, 1970); Karen LaChance, *Valdez: A Brief Oral History* (Valdez, AK: Prince William Sound Community College, 1995); and survivor interviews.

168 **a staff member played the piano:** National Research Council, *Great Alaska Earthquake*, 7:346.

170 **They didn't have fishing in mind:** One account of the Ferriers' experience can be found in Chance, *Chronology of Physical Events.*

172 **the first plane ride of his life:** Interview with Gary Minish.

173 **Owen Meals had a suggestion:** LaChance, *Valdez.*

11. The Barnacle Line

176 **The northern acorn barnacle:** D. T. Anderson, *Barnacles: Structure, Function, Development and Evolution* (New York: Springer, 1993).

176 **he became obsessed with barnacles:** Rebecca Stott, *Darwin and the Barnacle: The Story of One Tiny Creature and History's Most Spectacular Scientific Breakthrough* (New York: W. W. Norton, 2004).

178 **the three had gone their separate ways:** George Plafker, from his unpublished field notebooks.

181 **one look at the shoreline:** Details of this fieldwork are from interviews with George Plafker.

182 **In plain, straightforward language:** Arthur Grantz, George Plafker and Reuben Kachadoorian, *Alaska's Good Friday Earthquake, March 27, 1964: A Preliminary Geologic Evaluation* (Washington, DC: US Geological Survey, 1964).

183 **They have an air bladder:** Ibid., 12.

185 **an ambitious research program:** Detailed in Wallace R. Hansen et al., *The Alaska Earthquake of March 27, 1964: Field Investigations and Reconstruction Effort,* USGS Professional Paper 541 (Washington, DC: US Geological Survey, 1966).

12. Rebuilding

193 **it managed to arrive in Governor Bill Egan's office:** This letter and the others mentioned are in the University of Alaska–Fairbanks' Alaska and Polar Regions Collections.

194 **an early off-the-cuff estimate:** The governor's estimate was front-page news in the *Anchorage Daily Times* and other newspapers across the country the day after the quake.

194 **estimates of the disaster's cost had declined:** The estimate of $310 million is roughly equivalent to $2.4 billion in 2017 dollars.

195 **was at work almost immediately:** For an overview of the Army Corps of Engineers' work, see Patrick M. Coullahan and Allan D. Lucht, "Good Friday, 1964: The Great Alaskan Earthquake," *Military Engineer* (Society of American Military Engineers), n.d., http://themilitaryengineer.com/index.php/staging/item/307-good-friday-1964-the-great-alaskan-earthquake.

198 **Things didn't go exactly as planned:** National Research Council, *The Great Alaska Earthquake of 1964*, vol. 7, *Human Ecology* (Washington, DC: National Academy of Sciences, 1970), 138.

199 **Tatitlek residents would make this clear:** Unpublished interview with Mary Kompkoff by employee of Alaska Department of Fish and Game.

200 **The gift had made quite a splash:** The donation was reported in the *Bridgeport Post* (CT), among other newspapers.

201 **There was some grumbling about this idea:** For more on the relocation of Valdez, see National Research Council, *The Great Alaska Earthquake of 1964,* vol. 6, *Engineering* (Washington, DC: National Academy of Sciences, 1973).

201 **The planner, thirty-three-year-old Paul Finfer:** "After the Earthquake," *1982–83 Illinois Institute of Technology in the New Era* 4. See also Henry Saeman, "Quake Helpful, but Not Essential to City Plan," *Dayton Daily News,* August 31, 1966.

202 **One person suggested a burlesque hall:** "Citizen Questionnaire for Residential Section of Mineral Creek Townsite, Valdez, Alaska," July 1964, Valdez Museum and Historical Archive.

203 **As one resident put it later:** Karen LaChance, *Valdez: A Brief Oral History* (Valdez, AK: Prince William Sound Community College, 1995), 72.

13. Deep Thinking

208 **the subsidence was "secondary":** George Plafker and L. R. Mayo, "Tectonic Deformation, Subaqueous Slides and Destructive Waves," US Geological Survey open-file report, 1965.

208 **had moved about sixty feet:** National Research Council, *The Great Alaska Earthquake of 1964,* vol. 2, *Seismology and Geodesy* (Washington, DC: National Academy of Sciences, 1973), 428.

210 **a "rider" to Hess's hypothesis:** Vine used the term in print and in an undated lecture available on YouTube: "Fred Vine Explaining Paleomagnetic Reversals," uploaded May 12, 2008, youtu.be/CRx66ZpEhOg.

210 **Vine thought it remarkable:** Frederick J. Vine, "Reversals of Fortune," in *Plate Tectonics: An Insider's History of the Modern Theory of the Earth,* ed. Naomi Oreskes (Boulder, CO: Westview Press, 2001), 49.

211 **what was called polar wandering:** A thorough look at the development of plate tectonics, including work on understanding magnetic reversals, is William Glen, *The Road to Jaramillo: Critical Years of the Revolution in Earth Science* (Stanford, CA: Stanford University Press, 1982).

212 **the Vine-Matthews-Morley hypothesis:** Lawrence W. Morley describes his work in "The Zebra Pattern," in Oreskes, *Plate Tectonics,* 67–85.

214 **"You don't believe all this rubbish, do you?":** Edward C. Bullard, *William Maurice Ewing, 1906–1974* (Washington, DC: National Academy of Sciences, 1975).

214 **Cox had dropped out:** Konrad B. Krauskopf, *Allan V. Cox, 1926–1987* (Washington, DC: National Academy of Sciences, 1988).

214 **inadvertently ruining Doell's summer:** Interview with George Plafker.

216 **like him, a native New Yorker:** A short biography of Press is "Frank Press, 1981–93, NAS President," National Academy of Sciences Online, n.d., www.nasonline.org/about-nas/history/highlights/frank-press.html.

219 **Press's paper:** Frank Press and David Jackson, "Vertical Extent of Faulting and Elastic Strain Release," *Science* 147 (February 19, 1965): 867–968.

220 **"these measurements showed no abrupt changes":** George Plafker, "Tectonic Deformation Associated with the 1964 Alaskan Earthquake," *Science* 148 (June 25, 1965): 1675–87.

14. Acceptance

224 **several papers by Lamont researchers:** Among those publishing papers at the time were Walter Pitman, James Heirtzler and Xavier Le Pichon.

225 **Plafker's talk:** George Plafker, "Possible Evidence for Downward-Directed Mantle Convection Beneath the Eastern End of the Aleutian Arc," *American Geophysical Union Transactions* 48, no. 1 (1967): 218.

225 **Press approached him:** Interview with George Plafker.

226 **struck about one hundred miles offshore:** George Plafker and J. C. Savage, "Mechanism of the Chilean Earthquakes of May 21 and May 22, 1960," *Bulletin of the Geological Society of America* 81, no. 4 (1970): 1001–30.

227 **a young geophysicist from Princeton:** Background on Morgan's talk is from Xavier Le Pichon, "Introduction to the Publication of the Extended Outline of Jason Morgan's April 17, 1967, American Geophysical

Union Paper on 'Rises, Trenches, Great Faults and Crustal Blocks,'" *Tectonophysics* 187, nos. 1–3 (February 1991): 1–5.

228 **far greater knowledge of earthquake hazards:** For a concise discussion of the lasting impact of the 1964 quake, see Thomas M. Brocher et al., *The 1964 Great Alaska Earthquake and Tsunamis: A Modern Perspective and Enduring Legacies* (Menlo Park, CA: US Geological Survey, 2014).

230 **the last major earthquake:** For more on the potential for a Cascadia earthquake, see Kathryn Schulz, "The Really Big One," *New Yorker*, July 20, 2015.

15. Epilogue

232 **He returned to Guatemala:** George Plafker, "Tectonic Aspects of the Guatemala Earthquake of 4 February 1976," *Science*, n.s., 193, no. 4259 (September 24, 1976): 1201–8.

232 **a landslide caused by the quake:** George Plafker et al., "Geological Aspects of the May 31, 1979, Peru Earthquake," *Seismological Society of America Bulletin* 61, no. 3 (1979): 178–79.

235 **that resolve began to weaken:** Description of the effort to establish a new village are from Donald R. Poling, comp., *Chenega Diaries: Stories and Voices of Our Past* (n.p.: Chenega Corporation, 2011); William E. Simeone and Rita Miraglia, *An Ethnography of Chenega Bay and Tatitlek, Alaska* (Anchorage: Alaska Department of Fish and Game, 2000); and newspaper articles, including Helen Gillette, "After 16 Years, Still Rebuilding," *Anchorage Daily Times*, March 27, 1980.

235 **the need to do something about natives' claims:** A good short discussion of the Native Claims Act can be found in Claus M. Naske and Herman E. Slotnick, *Alaska: A History*, 3rd ed. (Norman: University of Oklahoma Press, 2014). For a more detailed look at the pipeline decision, see Peter A. Coates, *Trans-Alaskan Pipeline Controversy: Technology, Conservation and the Frontier* (Fairbanks: University of Alaska Press, 1993).

236 **yet another disaster struck Alaska:** University of Alaska–Fairbanks's Project Jukebox has oral histories from the spill: "Exxon Valdez Oil

Spill," Project Jukebox, n.d., http://jukebox.uaf.edu/site7/exxonvaldez, accessed in May 2016.

236 **a new community in Chenega Bay:** See Simeone and Miraglia, *Ethnography of Chenega Bay.*

238 **He'd gotten on a boat:** Interview with Tom and Gloria McAlister.

238 **There was even a proposal:** R. K. Alman, *Development Plan Report: Valdez Historic Site* (Anchorage: Alaska Department of Natural Resources, 1967).

239 **an icebreaker-tanker:** For more on this effort, see Ross Coen, *Breaking Ice for Arctic Oil: The Epic Voyage of the SS* Manhattan *Through the Northwest Passage* (Fairbanks: University of Alaska Press, 2012).

240 **The pipeline jobs changed the town:** Interview with Gary Minish; also see Karen LaChance, *Valdez: A Brief Oral History* (Valdez, AK: Prince William Sound Community College, 1995).

ADDITIONAL SOURCES

Anderson, Orson L. *Plate Tectonics: Selected Papers from the Journal of Geophysical Research.* Washington, DC: American Geophysical Union, 1972.

Borneman, Walter R. *Alaska: Saga of a Bold Land.* New York: Harper, 2003.

Boylan, Janet. *The Day Trees Bent to the Ground: Stories from the '64 Alaska Earthquake.* Anchorage: Publication Consultants, 2004.

Brothers, D. S., P. J. Haeussler, L. Liberty, D. Finlayson, E. Geist, K. Labay and M. Byerly. "A Submarine Landslide Source for the Devastating 1964 Chenega Tsunami, Southern Alaska," *Earth and Planetary Science Letters* 438 (March 15, 2016): 112–21.

Carson, Barbara. "Chenega's School Teacher Leaving: No More Chenega," *Anchorage Daily News,* April 7, 1964.

Carver, Gary, and George Plafker. "Paleoseismicity and Neotectonics of the Aleutian Subduction Zone—An Overview," in *Active Tectonics and Seismic Potential of Alaska,* edited by J. T. Freymueller, P. J. Haeussler, R. L. Wesson and G. Ekström, Washington, DC: American Geophysical Union, 2008.

Coats, R. R. "Magma Type and Crustal Structure in the Aleutian Arc," in *The Crust of the Pacific Basin, Geophysical Monograph Series* 6 (1962): 92–109.

Coll, Steve. *Private Empire: ExxonMobil and American Power.* New York: Penguin Press, 2012.

Daley, E. Ellen, ed. *Guide to Alaska Geologic and Mineral Information.* Fairbanks: Division of Geological and Geophysical Surveys, 1998.

Darwin, Charles R. *Living Cirripedia, a monograph on the sub-class Cirripedia, with figures of all the species. The Lepadidæ or pedunculated cirripides.* vol. I. London: The Ray Society, 1851.

Davis, Nancy Y. "The Exxon Valdez Oil Spill," in *The Long Road to Recovery: Community Responses to Industrial Disasters,* edited by James K. Mitchell. Tokyo: United Nations University Press, 1996.

De Laguna, Frederica. *Chugach Prehistory: The Archeology of Prince William Sound, Alaska.* Seattle: University of Washington Press, 1956.

Eckel, Edwin B. *Effects of the Earthquake of March 27, 1964, on Air and Water Transport, Communications, and Utilities Systems in South-Central Alaska.* USGS Professional Paper 545-B. Washington, DC: US Geological Survey, 1967.

———. *The Alaska Earthquake, March 27, 1964: Lessons and Conclusions.* USGS Professional Paper 546. Washington, DC: US Geological Survey, 1970.

Ford, Corey. *Where the Sea Breaks Its Back: The Epic Story of Early Naturalist Georg Steller and the Russian Exploration of Alaska.* Portland, OR: Alaska Northwest Books, 2003.

Frisch, Wolfgang, Martin Meschede and Ronald C. Blakey. *Plate Tectonics: Continental Drift and Mountain Building.* New York: Springer, 2011.

Fuis, Gary S., Peter J. Haeussler and Brian F. Atwater. "A Tribute to George Plafker," *Quaternary Science Reviews* 113 (2015): 3–7.

Garfield, Brian. *Thousand-Mile War: World War II in Alaska and the Aleutians.* Chicago: University of Chicago Press, 1995.

Gariepy, Henry. *A Century of Service in Alaska, 1898–1998: The Story and Saga of the Salvation Army in "The Last Frontier."* Rancho Palos Verdes, CA: The Salvation Army, USA Western Territory, 1998.

Grant, Ulysses Sherman, and Daniel F. Higgins. *Reconnaissance of the Geology and Mineral Resources of Prince William Sound, Alaska.* US Geological Survey Bulletin 443. Washington, DC: Government Printing Office, 1910.

Greene, Mott T. *Geology in the Nineteenth Century: Changing Views of a Changing World.* Ithaca, NY: Cornell University Press, 1985.

Hansen, Wallace R. *Effects of the Earthquake of March 27, 1964, at Anchorage, Alaska.* USGS Professional Paper 542-A. Washington, DC: US Geological Survey, 1965.

Haskell, William B. *Two Years in the Klondike and Alaskan Gold Fields, 1895–1898: Thrilling Narrative of Life in the Gold Mines and Camps.* Fairbanks: University of Alaska Press, 1998.

Hassen, Harold. *The Effect of European and American Contact on the Chugach Eskimo of Prince William Sound, Alaska, 1741–1930.* Unpublished PhD dissertation, University of Wisconsin–Milwaukee, 1978.

Hastie, L. M., and J. C. Savage. "A Dislocation Model for the 1964 Earthquake," *Bulletin of the Seismological Society of America* 60 (1970): 1389–92.

Haynes, Terry L., and Robert J. Wolfe, eds. *Ecology, Harvest, and Use of Harbor Seals and Sea Lions: Interview Materials from Alaska Native Hunters.* Juneau: Alaska Department of Fish and Game, 1999.

Hough, Susan E. *Earthshaking Science: What We Know (and Don't Know) about Earthquakes.* Princeton, NJ: Princeton University Press, 2004.

Johnson, John F. C. *Chugach Legends: Stories and Photographs of the Chugach Region.* Anchorage: Chugach Alaska Corporation, 1984.

Johnson, Robert D. *Ernest Gruening and the American Dissenting Tradition.* Cambridge, MA: Harvard University Press, 1998.

Kachadoorian, Reuben. *Effects of the Earthquake of March 27, 1964, on the Alaska Highway System.* USGS Professional Paper 545-C. Washington, DC: US Geological Survey, 1968.

———. *Effects of the Earthquake of March 27, 1964, at Whittier, Alaska.* USGS Professional Paper 542-B. Washington, DC: US Geological Survey, 1965.

Kirkby, M. J., and Anne V. Kirkby. *Erosion and Deposition on a Beach Raised by the 1964 Earthquake, Montague Island, Alaska.* USGS Professional Paper 543-H. Washington, DC: US Geological Survey, 1969.

Kizzia, Tom. "Lost Village Reborn 20 Years After Quake," *Anchorage Daily News,* August 5, 1984.

Knox, Robert G. "Broken Homes and Broken Lives in the Big City," *Juneau Alaska Empire,* July 27, 1964.

Lander, James. F. *Tsunamis Affecting Alaska, 1737–1996.* Washington, DC: Department of Commerce, 1996.

Larson, Lee. "Things Moving—Red Cross Gives Guns, New Boats," *Anchorage Daily News,* April 17, 1964.

Leask, Linda, Mary Killorin and Stephanie Martin. *Trends in Alaska's People and Economy.* Brochure prepared for Institute of Social and Economic Research, University of Alaska–Anchorage, 2001.

Lemke, Richard. *Effects of the Earthquake of March 27, 1964, at Seward, Alaska.* USGS Professional Paper 542-E. Washington, DC: US Geological Survey, 1967.

Lisle, Richard J., Peter Brabham and John Barnes. *Basic Geological Mapping.* Oxford: John Wiley and Sons, 2011.

Lyell, Charles. *Principles of Geology* (abridged edition). New York: Penguin Classics, 1998.

McClure, Val. *The Anchorage Chronicles—Eight Days of Disaster.* Unpublished account by Anchorage Armed Services YMCA program director.

McCoy, Roger M. *Ending in Ice: The Revolutionary Idea and Tragic Expedition of Alfred Wegener.* New York: Oxford University Press, 2006.

McCulloch, David S., and Manuel G. Bonilla. *Effects of the Earthquake of March 27, 1964, on the Alaska Railroad.* USGS Professional Paper 545-D. Washington, DC: US Geological Survey, 1970.

McGarr, Arthur, and Robert C. Vorhis. *Seismic Seiches from the March 1964 Alaska Earthquake.* USGS Professional Paper 544-E. Washington, DC: US Geological Survey, 1968.

McGinniss, Joe. *Going to Extremes.* New York: Alfred A. Knopf, 1980.

McPhee, John. *Coming into the Country.* New York: Farrar, Straus and Giroux, 1977.

———. *Annals of the Former World.* New York: Farrar, Straus and Giroux, 2000.

Miraglia, Rita. "The Cultural and Behavioral Impact of the Exxon Valdez Oil Spill on the Native Peoples of Prince William Sound, Alaska," *Spill Science and Technology Bulletin* 7 (2002): 75-87.

National Research Council. *The Great Alaska Earthquake of 1964,* vol. 1, *Hydrology.* Washington, DC: National Academy of Sciences, 1968.

———. *The Great Alaska Earthquake of 1964,* vol. 3, *Biology.* Washington, DC: National Academy of Sciences, 1971.

———. *The Great Alaska Earthquake of 1964,* vol. 4, *Geology.* Washington, DC: National Academy of Sciences, 1971.

———. *The Great Alaska Earthquake of 1964,* vol. 8, *Summary and Recommendations.* Washington, DC: National Academy of Sciences, 1973.

———. *Living on an Active Earth: Perspectives on Earthquake Science.* Washington, DC: National Academies Press, 2003.

Oreskes, Naomi. *The Rejection of Continental Drift: Theory and Method in American Earth Science.* New York: Oxford University Press, 1999.

Pearson, Roger, and Marjorie Hermans, eds. *Alaska in Maps: A Thematic Atlas.* Fairbanks: University of Alaska–Fairbanks, 1998.

Plafker, George. "The Alaskan Earthquake of 1964 and Chilean Earthquake of 1960; Implications for Arc Tectonics and Tsunami Generation," *Journal of Geophysical Research* 77, no. 5 (1972): 901–25.

———. *Field notebooks,* 1964–65 (unpublished).

———. *Surface Faults on Montague Island Associated with the 1964 Alaska Earthquake.* USGS Professional Paper 543-G. Washington, DC: US Geological Survey, 1967.

———. *Tectonics of the March 27, 1964 Alaska Earthquake.* USGS Professional Paper 543-I. Washington, DC: US Geological Survey, 1969.

Plafker, George, and Reuben Kachadoorian. *Geologic Effects of the March 1964 Earthquake and Associated Seismic Sea Waves on Kodiak and Nearby Islands, Alaska.* USGS Professional Paper 543-D. Washington, DC: US Geological Survey, 1966.

Plafker, George, and Meyer Rubin. "Uplift History and Earthquake Recurrence as Deduced from Marine Terraces on Middleton Island, Alaska," in *Proceedings of Conference VI, Methodology for Identifying Seismic Gaps and Soon-to-Break Gaps.* US Geological Survey open-file report, 1978.

Plafker, G., R. Kachadoorian, E. B. Eckel and L. R. Mayo. *Effects of the Earthquake of March 27, 1964, on Various Communities.* USGS Professional Paper 542-G. Washington, DC: US Geological Survey, 1969.

Powers, Dennis M. *The Raging Sea: The Powerful Account of the Worst Tsunami in U.S. History.* Wellesley, MA: Sea Ventures Press, second ed., 2015.

Sandberg, Eric. *A History of Alaska Population Settlement.* Juneau: Alaska Department of Labor and Workforce Development, 2013.

"Starting Over," unbylined article, *Anchorage Daily News,* September 2, 1984.

Stowell, Harold. *Geology of Southeast Alaska: Rock and Ice in Motion.* Fairbanks: University of Alaska Press, 2006.

Stratton, Lee, and Evelyn B. Chisum. *Resource Use Patterns in Chenega, Western Prince William Sound: Chenega in the 1960s and Chenega Bay, 1984–1986.* Anchorage: Alaska Department of Fish and Game, 1986.

Sullivan, Walter. "Science: The Earth's Upheavals," *New York Times,* July 11, 1965.

———. "Origin of Alaska Quake Traced Over 950 Years by Scientist," *New York Times,* April 20, 1967.

Takeuchi, Hitoshi, Seiya Uyeda and Hiroo Kanamori. *Debate About the Earth: Approach to Geophysics Through Analysis of Continental Drift.* Translated by Keiko Kanamori. San Francisco: Freeman, Cooper, 1967.

Tower, Elizabeth A. *Alaska's Homegrown Governor: A Biography of William A. Egan.* Anchorage: Publication Consultants, 2003.

Van Brocklin, Thomas. *Historic Homes of Old Valdez.* Valdez, AK: Prince William Sound Books, 1987.

Vine, Frederick J. "The Continental Drift Debate," *Nature* 266 (March 3, 1977): 19–22.

Vine, Frederick J., and Drummond H. Matthews. "Magnetic Anomalies over Oceanic Ridges," *Nature* 199 (September 7, 1963): 947–49.

Vorhis, Robert C. *Hydrologic Effects of the Earthquake of March 27, 1964, Outside Alaska.* USGS Professional Paper 544-C. Washington, DC: US Geological Survey, 1967.

Wilson, Basil W., and Alf Torum. *The Tsunami of the Alaska Earthquake, 1964: Engineering Evaluation.* Coastal Engineering Research Center, US Army Corps of Engineers, 1968.

INDEX

ABOUT THE AUTHOR

HENRY FOUNTAIN has been a reporter and editor at the *New York Times* for two decades, writing about science for most of that time. From 1999 to 2009, he wrote Observatory, a weekly column in the Science Times section. He was an editor on the national news desk and the Sunday Week in Review and was one of the first editors of Circuits, the *Times*'s pioneering technology section. Prior to coming to the *Times,* Fountain worked at the *International Herald Tribune* in Paris, *New York Newsday* and the *Bridgeport Post* in Connecticut. He is a graduate of Yale University, where he majored in architecture. He and his family live near New York City.